the aquarium fish handbook

David Goodwin

This edition published by Magnet & Steel Ltd

Printed 2014

This book is distributed in the UK by
Magnet & Steel Ltd
Unit 6
Vale Business Park
Cowbridge
CF71 7PF

sales@magnetsteel.com

ISBN: 978-1-907337-72-7

Printed by Printworks Global Ltd., London/Hong Kong

contents

6

Introduction

Watching fish swimming around an aquarium is extremely calming and peaceful, especially after a stressful day. Having spent time and money designing and setting up an aquarium, it is very satisfying to watch the fruits of your labours flourish and possibly breed. Fish undoubtedly have a soothing and therapeutic effect upon the human psyche.

Introduction

A growing number of offices, hospitals and reception areas incorporate an aquarium in order to help clients relax before appointments.

Tropical fish make ideal pets for children. Unlike larger animals, such as cats and dogs, fish do not have to be taken for walks or require grooming. They have to be fed, and the tank has to be cleaned, but this is a comparatively gentle introduction to the responsibilities of caring for a living creature.

The expense of keeping tropical fish depends very much on the size of aquarium you choose. The biggest cost is the initial purchase of the aquarium, cabinet and equipment needed to run it efficiently. Buy the very best that you can afford. General running costs are quite low: some everyday foods can be prepared for consumption by fish, and some live foods, such as daphnia or cyclops, can be collected from local ponds or streams. A few nicely shaped, large pebbles can be collected during a walk on a beach. Always remember to boil, scrub and thoroughly sterilise them before placing them in your aquarium. Thin pieces of rock or slate can be bonded together using aquarium silicone sealant to make caves or hiding places for your fish. (Tip: do not use bathroom sealants in your aquarium. They have various colour pigments in them, which are toxic to your fish.) Short pieces of plastic pipe can be bonded together and then coated in gravel so that they will blend into the aquarium set-up, again to make hideaways for your fish.

Owning an aquarium can be a pastime that is calming and restful.

General advice

First, find a well-stocked specialist shop with knowledgeable and helpful staff. It is worth shopping around to compare both prices and stocks. By frequenting one shop regularly, you will build up a good relationship with the staff and receive better service.

Check that the shop maintains its fish tanks properly and never buy fish that appear to be ailing or unhealthy.

Do not be afraid to ask questions about fish, plants or equipment.

Carefully plan the size of the aquarium you want and work out how much it will all cost. (Every fish-keeper that I know has wanted a larger aquarium within six months).

Children will probably be keen to help install the new aquarium, but please supervise them closely, particularly if chemicals or electrical equipment are involved.

Red & White Fantail

Male Fighter – variation

Water quality

Water quality varies across the country; the level of acidity or alkalinity, and thus the pH value, depends on the water's source. On a pH scale, 0 is the acidic end of the range and 10 the alkaline. A reading between 6.8 and 7.0 is regarded as neutral. Aquarium fish must have water of the correct pH. If it is incorrect, the fish may be unable to breed, and in some instances, if it is too low they may even die. Most of the river Amazon in South America is on the acidic side of neutral, for example, whereas the Rift Valley lakes in Africa have very hard water, with pH readings never lower than 8. Always check the water requirements of the fish that you wish to purchase and do not put them through any unnecessary pain or stress. As a guideline for a general community aquarium, a pH level somewhere between 6.6 and 7.5 should adequate.

In some areas, the water is referred to as 'soft', meaning that it is faintly acidic because it lacks calcium and various other minerals. If soap lathers easily in your tap water, then you probably live in an area of soft water.

'Hard' water is drawn through substrate in mainly limestone, sandstone or chalk areas, which gives it an alkaline pH value. Tap water may appear chalky or cloudy because of the calcium content. Soap does not lather so readily. There are two types of hardness: temporary hardness is caused by the presence of calcium and magnesium bicarbonates, which can be removed. Permanent hardness derives from a number of sulphates, such as calcium and magnesium, in the water, as well as the chlorides that are used for the purification of the water supply.

Testing and altering the pH value

There are a number of reasonably priced dip strips, liquid chemicals, tablets and electronic meters available to measure pH values. A local supplier will always recommend items, and will also be able to give you a good idea of pH levels in your area. Boiling the water and allowing it to cool right down before putting it into your aquarium will very easily remove temporary hardness. Under no circumstances should you place boiling or hot water into your aquarium. Permanent hardness can be removed, but is not advisable for tropical fish. If you need to raise the pH, add sodium bicarbonate in small amounts and constantly check until the reading is correct. Acid buffers can also be purchased

Toxins

Toxins can build up in an aquarium from decaying waste, overcrowding and inefficient filtration. Nitrites and ammonia can be highly toxic to fish, even at very low levels. Nitrites are created by the lack of oxygen and are a highly poisonous toxin, even if they are present in a weak concentration.

A combination of high levels of ammonium and a pH reading of over 6.8 helps create ammonia in the aquarium. Low ammonium levels on their own are not too much of a problem, but will increase if there is decaying waste matter, combined with too many fish in the aquarium and overfeeding. Very low levels of ammonia are highly toxic.

These toxins can be avoided very easily by keeping the numbers of fish consistent with the size of the aquarium, and ensuring that the filtration system is efficient. Feed the fish little and often. Change the water regularly, replacing 25 per cent a month, or preferably 10 ten per cent a week Test kits are readily available, and you should do tests on a regular basis, weekly being best. If the reading is too high, replace 25 per cent of the water immediately and test 30 minutes later. If the reading is still higher than normal, change a further 25 per cent of the water. Check the aquarium for dead and decaying fish and get rid of any.

Fish can build up resistance to high levels of nitrates and ammonia. The only way to detect these toxins is to test the water regularly.

Aquariums can be filled with water from a number of different sources. Tap water is chemically treated to make it suitable for human consumption and should be dechlorinated before it is used in an aquarium. Alternatively, it can be aerated in a holding container for 24 hours, which also ensures that the water adjusts to the same temperature as the aquarium.

Distilled water is a very expensive way of filling an aquarium and lacks many base minerals required by fish. It can be used in small percentages in breeding tanks for certain fish.

Rain water is a useful source if you live in a rural area, but urban

Marliers Julie

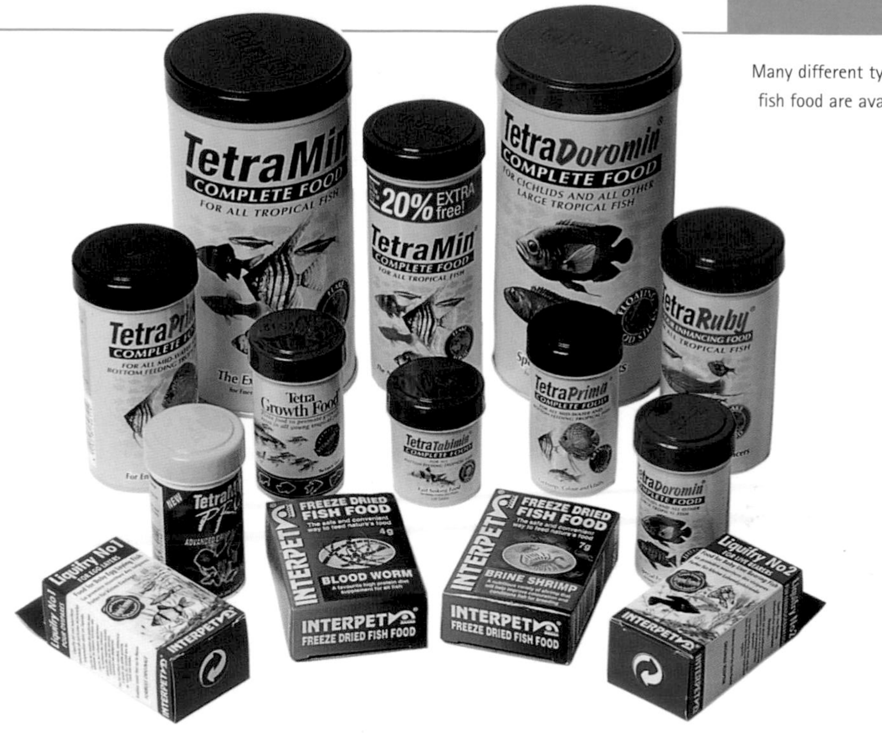

Many different types of fish food are available.

rain will inevitably contain too many toxins. Collect rainwater in a wooden or plastic butt. Always allow it to settle for two or three hours, and never drain from the bottom, but from about 6" (15.24 cm) off the base.

Pond water is another valuable source, but will contain unknown organisms and harmful pests and must be filtered very finely.

Reverse osmosis water is filtered through a semi-permeable membrane that filters out up to 90 per cent of the base minerals and impurities in a water supply. It can be excellent when used for breeding and keeping certain fish. The filtration process will reduce the level of the pH and must therefore be tested.

Foods and feeding

There are many different types of food available. Many fish need specific vitamins and other nutrients on a daily basis, which can be provided by a varied diet.

Buy only the best-quality flake foods available. The range is now tremendous, with some feeds including additives, such as spirulina, for algae-eating fish, colour-enhancers to bring out the best colours in your fish and low-level medications to maintain their general health. Flake foods are available in different sizes for larger fish.

Begin with a standard staple food and then learn about the individual requirements of your

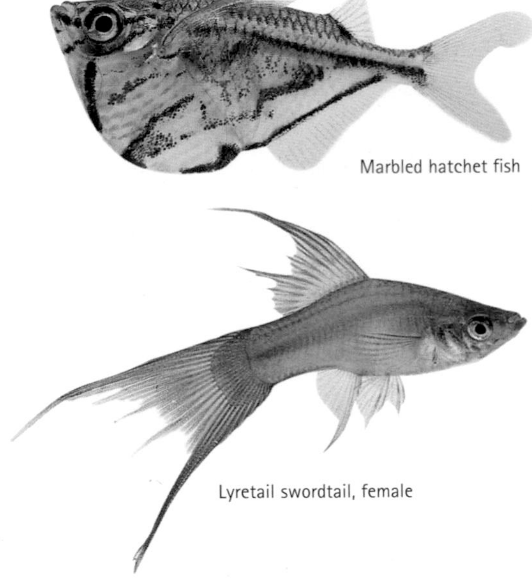
Marbled hatchet fish

Lyretail swordtail, female

fish. Always keep three or four different types of food available. Do not just feed flake food, as fish prefer a varied diet.

Dried food is also available as pellets, tablets or freeze-dried. Pellet food is mainly for surface-feeders and bottom-feeders; there are different sizes, depending on the size of your fish. Some of the pellets contain air to make them float and are ideal for fish that have an upturned mouth to feed from the surface. Other pellets sink quite quickly to the bottom and, as they become saturated, will soften up for the fish to feed on. Tablet food will also normally sink to the bottom, making it ideal for fish like corydoras and many other catfish. Freeze-dried foods are very useful if you are unable to obtain a ready supply of live foods, such as daphnia, bloodworms, tubifex and black mosquito larvae.

Live foods

Unfortunately, some aquatic shops do not have enough sales to warrant stocking live foods, so it may be hard to find a supplier. There are many live foods available, such as tubifex, daphnia, bloodworms, glassworms, black mosquito larvae, brine shrimp and river shrimp. Only ever buy as much as your fish will want for one day unless you are prepared to maintain these foods. You will have to refrigerate some of them. Adding this type of food to the aquarium will provoke a feeding frenzy among the fish, who will hunt for it after it has all been devoured. Always rinse through this type of food with fresh water before placing it in the aquarium. Beware of glassworm: although very small, it can be predatory if placed in an aquarium with fry.

Frozen foods

Frozen foods is probably the most convenient way to feed your fish. The many varieties of frozen food now available will feed just about any type of fish. Multi-menu packs contain four or five different types of food in a handy blister pack, which means that one small cube can be used at a time.

Kitchen food

Fish can also eat certain vegetables and meat in small portions. Vegetable- and algae-eating fish love cooked garden peas. The large, outer leaves of a soft lettuce can be used, too. Rinse them under cold water, then dip them into boiling water for a couple of seconds. By attaching a small weight to the bottom of the leaf, it will sink to the bottom of the aquarium for your fish to relish. Carnivorous fish enjoy beef heart. Freeze a small quantity and, when it is solid, grate off just enough for one feed. Defrost it and then feed to the fish

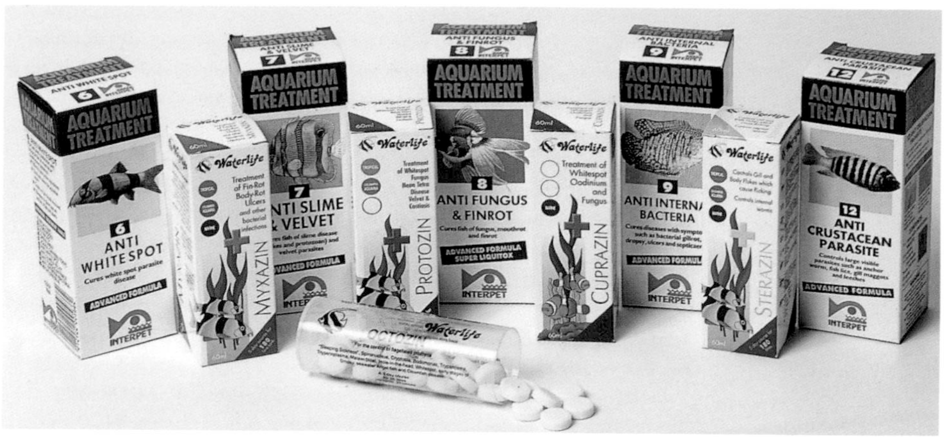

Proprietary treatments are available to combat the most common fish ailments.

Other foods

Earthworms are an excellent food for larger carnivorous fish. They must be cleaned before feeding by placing them in a very shallow saucer of milk for a couple of minutes and then draining them on a paper towel for a further five minutes. They can then be chopped or diced up before being added to an aquarium. Mealworms are another very good food, but only for larger fish. Use only a few at a time, as they are a very fatty food and will foul the aquarium water very quickly if they are not eaten. Live crickets are a very tasty meal for really good-sized fish.

Feed fish at the same time each day: the fish will get to know the feeder. It might be sensible to restrict feeding to just one person, which will help to ensure that the fish are not overfed. The feeder will also know exactly what variety of food the fish have eaten. If there is still food left in the tank after 10 minutes, you are overfeeding. Scoop out the food to avoid fouling the tank.

Diseases

To reduce the risk of any type of disease in the main tank, you should acquire a small quarantine tank. Any new fish should be placed in it for about 10 days, or until you are sure that the fish is healthy. It can then be moved to the main aquarium. If a fish from the main aquarium requires treatment, you can use the quarantine tank to treat the problem. A quarantine tank only needs to be very basic, being filtered and aerated, possibly with an internal power filter to do both jobs. and then

Dwarf Gourami – female

heated to the correct temperature. Other than this, the tank can be bare.

There are basically three different types of disease: bacterial, parasitic and viral. Fungal problems are usually a secondary disease.

White spot – ichthyopthirius

This is by far the most common problem in an aquarium, and is normally caused by shock or stress. Simply catching or transporting fish, banging on the front of the aquarium, or large changes in temperature between one aquarium and another, could bring this on. The symptoms are that the fish will start flicking itself on any item in the aquarium; tiny, pinhead-sized white spots then appear on the body and fins. If left untreated, the fish can die. There are many proprietary treatments available to cure white spot. Alternatively, increase the temperature of the heater thermostat by 10° for a period of 24 hours and then turn the control back down to the normal setting, which will normally cure this problem. Do not put hot water or cold water into the aquarium to increase or decrease the temperature. Let the heater thermostat heat it up and then allow it to cool back down to its normal temperature naturally.

Velvet - oodinium

This is similar to white spot, but is normally restricted to the sides and crown of the back. The spots are very fine, like dust, and are gold in colour. It is easier to see while a fish is swimming and turning in the light. It is very easy to cure with a proprietary medication. If you have a quarantine tank, place the fish in it and treat with 1oz (28g) of aquarium salt per gallon (3.8l) of water, which should cure it within two to three days. If it proves stubborn, add an additional 1oz (28g) of aquarium salt per gallon (3.8l) to the aquarium. After another day or two, you should find that the fish is cured.

Ulceration

Small lumps appear underneath the skin and slowly grow, like a boil. The skin will break open, leaving unsightly wounds. The best cure is to place the fish in a quarantine tank in clean water, raise the temperature slightly and then ask an aquatic shopkeeper for their recommended medication. It will take time and patience to cure this problem.

Gill infections

This type of infection can be created by either bacterial or parasitic problems. Study the problem, note the symptoms and consult a local dealer, who will recommend some medication. A bacterial remedy will not cure a parasitic problem and *vice versa*.

Long Fin Rosy Barb

Gold marble sailfin molly, male.

Tip: if you keep mollies, especially black mollies, be aware that they require some salt in the water; if this is not present, they show small, white patches on their body, similar to white spot. Treating for white spot will not cure this. They must be placed in a salt bath for a short period of time and then be replaced in their tank.

Fungal problems

Spores of fungi are always present in aquariums, and healthy fish are usually unaffected by them. Unhealthy fish are more susceptible, however. Mouth fungus can be a problem in bottom-feeders in an aquarium with a crushed-gravel substrate. Body fungus can appear if the fish are roughly handled: their outer skin can be removed by abrasive handling in the net or by fighting other fish. If the skin's protective layer is damaged or removed, fungus spores will go straight to work. Careful handling can prevent this, as can the use of round gravel as a substrate.

Problems occur in the best-kept tanks. If your fish become ill, do not panic, but try to solve the problem as quickly as possible. Tip: wear plastic or medical gloves when treating the fish so that you do not damage your skin if you spill chemicals or treatments. Check the life span of any treatment: the life of some chemicals can be very short after opening.

Black sailfin molly, female.

Dispose of them carefully and correctly and buy replacements only when needed.

Make sure that the fish you purchase are swimming actively and look healthy, with no obvious sores or lesions. Although many diseases are easy to cure, if the fish that you wish to purchase has a problem, speak to the shopkeeper about it and let him or her cure the problem before you buy it. They may not always be aware of a problem and will welcome your noticing it.

The aquarium

For many years, metal-framed aquariums were the only type available, but they have now been superseded by all-glass or perspex aquariums. Plastic tanks are fine for hatching out fish eggs and holding them for a few days, but they scratch and mark very easily. Before purchasing an aquarium, decide where you want to place it, a decision that will govern the size. The larger, the better. It is easier to manage the quality of a larger volume of water. If disease strikes, it quickly becomes a dominant problem in a small volume of water, whereas it is only a minor problem in a larger area as you will see it spread and treat it at an earlier stage.

TYPES OF AQUARIUM

There are a number of differing types of aquarium available:

metal-framed, putty-lined (out of date and very old fashioned)

plastic, all-in-one moulded aquariums

all-glass aquariums with no framing, bonded using aquarium silicone sealant

all-glass aquariums with built-in filtration systems

all-in-one moulding of perspex (some include filtration systems)

Where to place the aquarium

Try to find a light spot out of bright sunlight. A great deal of natural light produces high levels of algal growth over just about every surface, including plants and gravel. Unless you have a lot of algae-eating fish, there will be constant

Tiger barb

Flag-tail corydoras

Sheepshead acara

cleaning chores. Do not place the aquarium in a very dark corner because you will barely be able to view the contents. If the aquarium is to be built into a wall unit, ensure that there is enough space for regular maintenance. Do not place it next to a radiator because of the fluctuation in temperature.

Design

Aquariums are now available in all shapes and sizes, with the design limited only by your imagination. You can purchase the aquarium on its own and put it on a stand or you can buy one in a cabinet; they are produced in many different finishes, one of which should match the décor of your home. The aquarium is going to be a piece of furniture in your house, so take your time when deciding what you would really like and shop around for it.

Setting up the aquarium

Before you actually get to the point of setting up the aquarium, decide what type of gravel and rocks to use. Gravel ranges from coloured, spa and quartz to pea. Pea gravel is the best, as it allows plenty of water to flow through it for filtration, has no sharp edges, so the fish will not damage their mouths when feeding, and is similar in size to the gravel that the fish swim over in the wild.

Carefully consider the size and type of aquarium that you want.

SAFETY TIPS

Always ensure that the glass edges of the aquarium are smoothed down, including the corners, or are covered in tape or plastic edging.

Ensure that the power supply is turned off when working on, installing or replacing electrical appliances, such as the heater, pump or lighting units.

Calculate the full weight of the aquarium and make sure that your flooring will accept this weight. Do not site your aquarium along the length of the flooring joists; position it across the joists.

Ensure that the cabinet or stand is strong enough to support the weight of the filled aquarium.

Sit the aquarium on a sheet of polystyrene on the stand to absorb any slight imperfections.

Wash gravel thoroughly before placing it into the aquarium to get rid of the dust. Check that rocks are suitable by carrying out the 'vinegar test'. If you drip a few drops of vinegar onto the rock and it fizzes or bubbles, do not place it in the aquarium. If it just rolls off and without reacting, wash the rock thoroughly ready for use.

Roughly sketch a plan of the layout of plants and décor within the tank.

If you are going to place an aquarium background picture on the rear of the tank, do it now, because you will not be able to move the tank when it is full.

Ensure that the stand or cabinet is safe and stable and then lay a sheet of 1.5in. (13mm) polystyrene on it and it cut to the same base size as the aquarium. Carefully lift and sit the aquarium on this. Always get assistance with

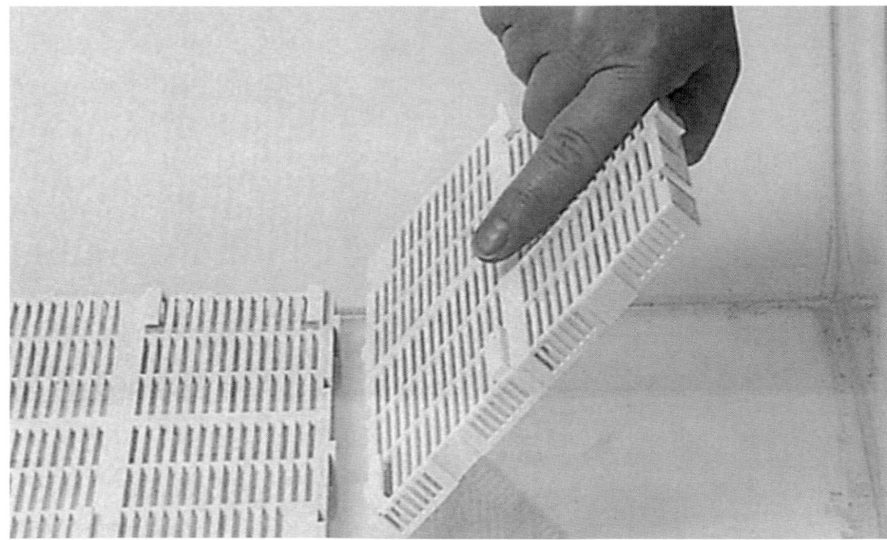

Figure 1

Figure 2

Figure 3

Figure 4

Figure 5

Figure 6

this so that you do not injure yourself.

Place the under-gravel filter plate inside the aquarium (Figure 1), ensuring that there is nothing under the edges, otherwise a secure seal will not be made. You can, if you wish, use aquarium silicone sealant to bond the plate to the base, but it will always be difficult to remove once you have done this. You will also have to wait about 24 hours for the silicone sealant to cure.

Put the uplift tube in place, ensuring that the top outlet will be 2-3in. (5-7cm) below the level of the water surface when the aquarium is filled (Figure 2). Pour the gravel on top of the filter, putting it onto the centre of the filter first (Figure 3). When all of the gravel is in the aquarium, push it out to the edges. A minimum depth of 3in. (8cm) of gravel is necessary for it to be an efficient biological filter. As a general guideline, 15lb (7kg) of gravel will be required per square foot (30cm square) of base area. More can be used if you wish. It is normally cheaper to buy gravel in 55 (25kg) sacks.

When you start to fill the tank, put a plate, right side facing upwards, on the gravel in the centre of the aquarium to stop the gravel from being thrown everywhere, especially if you have contoured it (Figure 4). Continue to add water until the tank is half full and then remove the plate.

Next to come is the heater/thermostat unit (Figure.5). Set the thermometer to 75˚F (24°C). Feed the supply cable from inside the tank up through the corner cut-outs in the stress bars. Using the suckers provided, stick it to the rear of the aquarium at an angle of 45°, with the thermostat control to the top. Do not put any part of this unit in the gravel, as it could

Figure 7

Figure 8

overheat and become a danger. If the gravel does start to build up against it, always clear it away. Now position the thermometer. If it is the stick-on type,stick it on the outside of the front glass, in one of the top corners. If it is a glass thermometer, then use the rubber sucker to stick it to the inside front glass, in either of the top corners.

Secure the light fittings in the aquarium lid and fix the fluorescent tube into the end caps. Connect the airline to the air pump and gently push the end of the airline into the top of the uplift tube, Push it all the way down into the tank and raise it 1in. (2.5cm) from the bottom.

Make sure that all of the electrical connections are correct, but do not connect the power supply yet. Remind yourself of the projected design and check that everything is ready to install. Now position the rocks or bogwood (Figure 6). Move them around until you are happy with the design. It is easier to do this while the water level is low.

Once everything is in place, put the plate back into it and continue to add water to the tank until it is full.When then tank is full, carefully lift out the plate, wipe off any water spillage from the stress bars and lay the condensation cover in place. Put the hood on the top of the tank, ensuring that no cables are trapped. Plug in the power supply and switch it

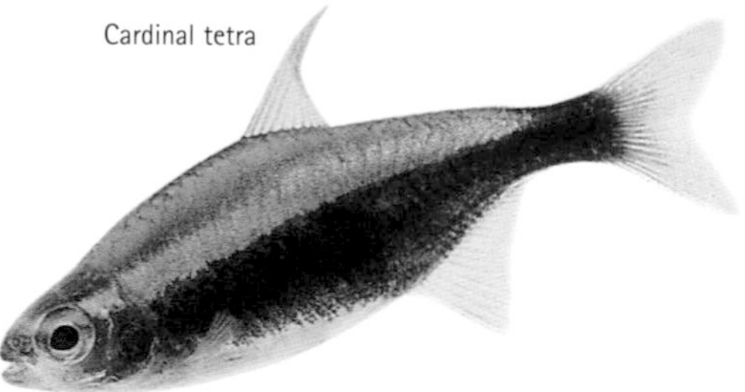

Cardinal tetra

on. If everything has been connected correctly, the light should be on, the heater should be working and the pump should be blowing bubbles out of the uplift tube. Now the plants can be added (Figure 7).

Place the plants in position and then bury their bases in the gravel, ensuring that the root systems are below it. Use plant pellets to feed the plant roots, especially when the tank is new, by pushing them into the gravel at the side of the plants.

Aftercare

It will take 24 to 36 hours for the temperature to stabilise, and it will help slightly if you leave the light on for this period. Once this has been done, add an aquarium conditioner to improve the quality of the water and allow another 24 hours for it to work. Any cloudiness in the water should vanish and the aquarium should be crystal clear. If it is still cloudy, turn off the power supply and change 50 per cent of the water. Turn the power back on and leave for a further 24 hours. When all is well, you are ready to purchase some fish (Figure 8).

Aquatic plants

There are many different types of plant sold to the aquatic hobbyist. Some are truly aquatic and many others are just bog plants that are grown and sold for their colour and decorative value. Unfortunately, bog or marsh plants do not survive for very long when fully immersed in an aquarium. The plants shown on the next

Wisteria.

Straight Vallis.

Bronze Bacopa.

Indian Fern.

Red Hygro.

Baby Tears.

Moneywort.

pages are capable of surviving in a fully submersed situation.

Plants are valuable as decoration, resting areas for the fish, spawning areas and places for the fish to hide when they feel threatened. The plants will also absorb carbon dioxide and give off oxygen into the water.

Natural light is obviously the best source of light, but if the aquarium is in a dark area of a room, then artificial lighting has to be used. A combination of differing light tubes should be used, because each different tube used gives light at different parts of the spectrum. If possible, use a timer to set the tubes to light up and turn off at differing times to create a 'sunrise' and 'sunset' for the fish and plants. Change the tubes at regular intervals, approximately every six to nine months.

Generally, the darker the leaf of the plant, the less light it requires. As an example, *Anubias nana* has a very dark leaf and could be planted in the shaded area of a piece of bogwood or rock, whereas *Cabomba* is a much lighter colour and can therefore be planted in the open.

A number of factors will govern whether the plants in an aquarium flourish:

amount and type of light available and for how long

type of filtration used

whether aquarium plant feed and nutrients are used

type of fish in the aquarium

layout of plants in the aquarium

Twisted Vallis.

Elodea.

Hygro Polysperma.

Cabomba.

Amazon Swords.

Teacup Stingray

Filtration is a personal choice, but undergravel filters are excellent for plant growth as long as they are regularly maintained. Plants need to have the mulm that settles around the base of their stems and in the gravel cleaned away on a regular basis so that the root systems can continue to feed on nutrients and bacteria.

Aquarium plants benefit from feeding in the same way as garden or house plants. There are liquid, tablet and pellet forms of plant feed to place in aquariums quite safely. Do not use houseplant feeds, because they might poison the fish. The latest and most efficient way to care for plants is to install a CO_2 system in your aquarium.

The choice of fish to go into a planted aquarium is also very important. Some fish eat plants or pull them out of the substrate, so choose the fish with care if you want a well-planted aquarium. Remember, however, that the plants are installed mainly for the benefit of the fish and not *vice versa*.

Fish Anatomy

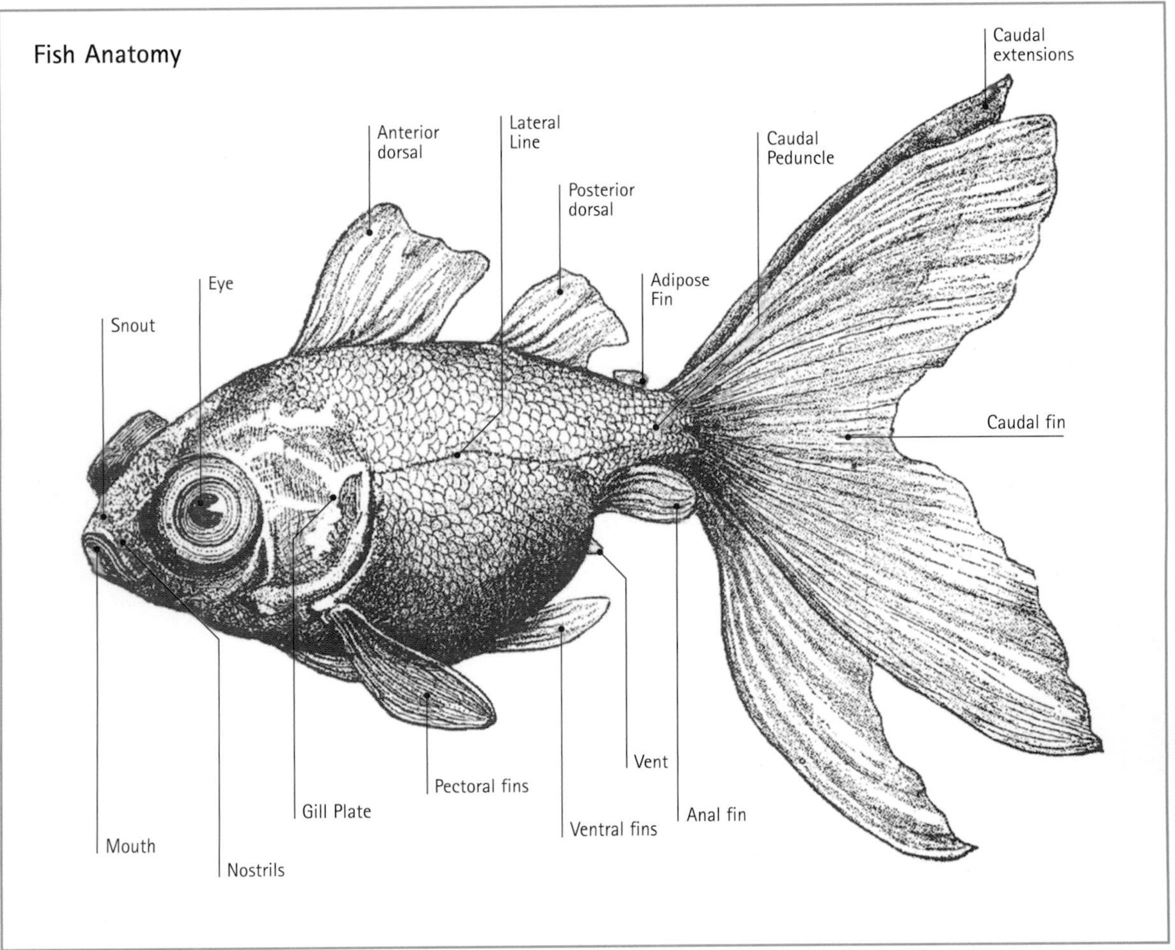

General information

This book is intended to provide a general introduction to everyday fish-keeping. We have tried to avoid using too much technical information, but there are certain facts that are vital to keeping healthy fish.

Since the 18th century, scientists have classified all living things by genus (or family) and species. We have noted both the common and Latin names of the fish in this book. The Latin name consists of the genus, followed by the species; the red-tailed black shark, for example, is *Labeo bicolor*, *Labeo* being the genus and *bicolor* the species.

Neon Blue Dwarf Gourami – male

CALCULATIONS AND IMPERIAL TO METRIC CONVERSIONS

To calculate the surface area of the aquarium:
multiply the length x width = surface area.
(In an unfiltered aquarium, you should allow 12" (77cm) square of surface area to 1" (2.5cm) in length of fish as a stocking level.)

To calculate the cubic volume of the aquarium:
multiply the length x width x height = cubic volume.
To calculate the water volume of the aquarium:
imperial measurement – multiply the length x width x height x 6.23
(do not forget to deduct from your measurements the depth of the substrate).

To calculate the weight of the aquarium, including the water:
imperial measurement – multiply the length x width x height x 6.23 x 10.

To convert centigrade into fahrenheit:
multiply by 9/5ths and add 32.

To convert fahrenheit into centigrade:
subtract 32 and multiply by 5/9th.

1 imperial gallon of water weighs 10lb (4.5kg).

1 imperial gallon of water = 4.54 litres.

1 litre of water weighs 1kg (2.2lb).

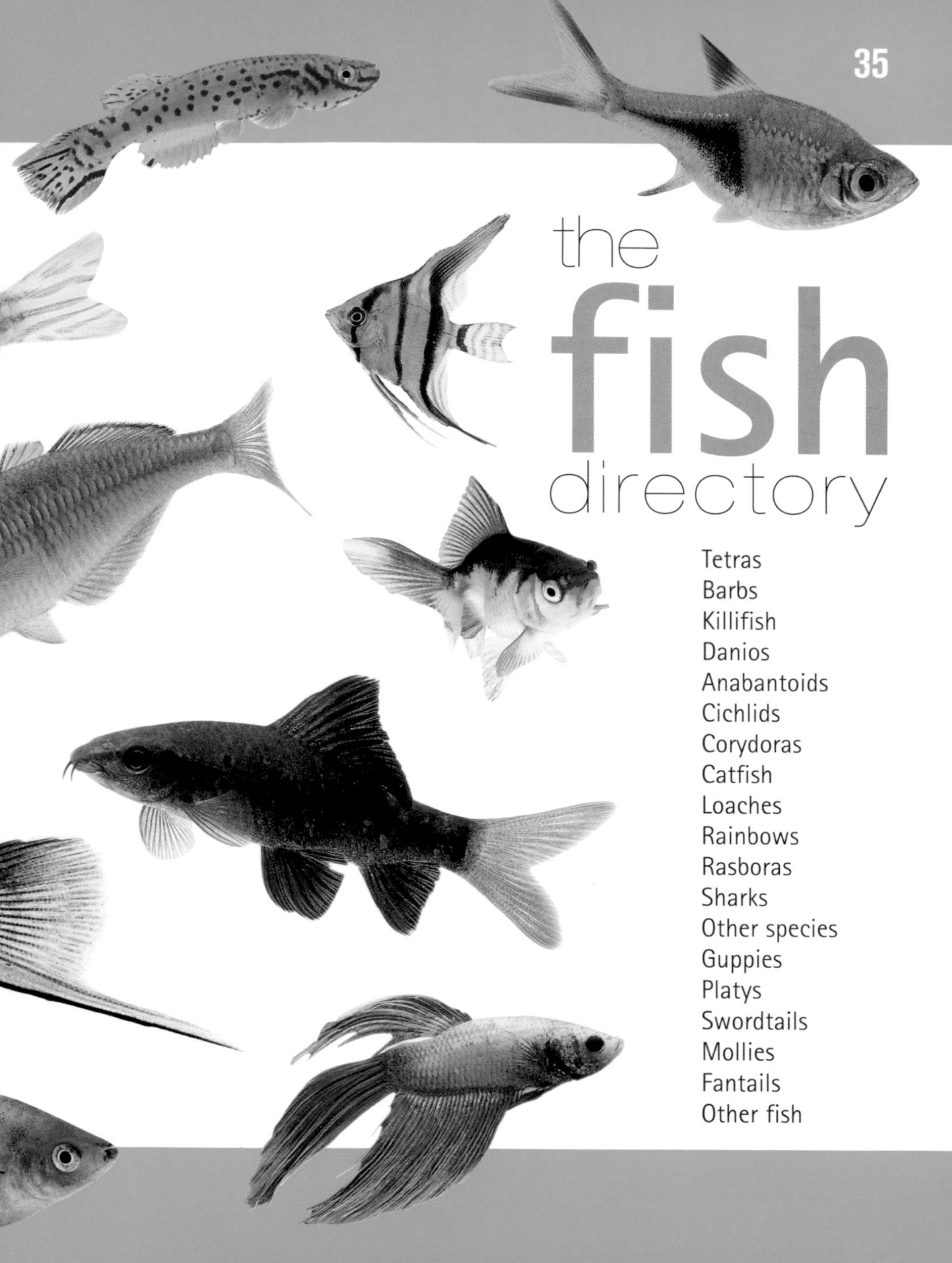

the fish directory

- Tetras
- Barbs
- Killifish
- Danios
- Anabantoids
- Cichlids
- Corydoras
- Catfish
- Loaches
- Rainbows
- Rasboras
- Sharks
- Other species
- Guppies
- Platys
- Swordtails
- Mollies
- Fantails
- Other fish

Neon tetra / *Paracheirodon innesi*

ORIGIN
South America

TEMPERATURE RANGE
22–26°C (72–79°F)

COMMUNITY
excellent

ADULT SIZE
female 4cm (1.5in.)
male 4cm (1.5in.)

DIET
all foods

EASE OF KEEPING
8/10

PH RANGE
6.8–7.5

This is one of the three most popular aquarium fish in the world. Best kept in shoals of 10 or more, they will show up exceptionally well in an aquarium, constantly shoaling together. Do not put them into a new aquarium,but wait 10 to 12 weeks until the bacteria in the filters has started to work.They can be a little touchy when first transported, so check for white spot within a couple of days and treat if necessary, but do not let this put you off purchasing them. Look after them well and they will give you much pleasure and live for a long time. Ensure that there are plenty of tall, fine plants in the aquarium (such as *Cabomba* or *Elodea*), so that if they feel threatened they can hide.The females tend to have much fuller bodies than the males.

Feeding

They will take all types of food, but do much better on a wide variety of dried, frozen and live foods. Make sure that their food is broken down small enough for them to eat.

Cardinal tetra / *Cheirodon axelrodii*

This is another extremely popular fish. Best kept in shoals of 10 or more, they have vivid red-and-blue coloration which shows up well. Unlike neons, the cardinals' lines of colour extend the full length of the body. Add cardinals to a new aquarium after approximately 10 to 12 weeks, when the bacteria in the filters is working. These fish can also be prone to white spot, so treat if necessary. Many high-quality cardinals are supplied by commercial breeders in the Czech Republic. Ensure that there are plenty of tall, fine plants in the aquarium (such as *Cabomba* or *Elodea*), so that if they feel threatened they can go for cover. The females tend to have a much fuller body than the males.

Feeding

They will take all types of food, but do much better on a wide variety of dried, frozen and live foods. Make sure that their food is broken down small enough for them to eat.

ORIGIN
South America

TEMPERATURE RANGE
22–26°C (72–79°F)

COMMUNITY
excellent

ADULT SIZE
female 4.5cm (1.75in.)
male 4.5cm (1.75in.)

DIET
all foods

EASE OF KEEPING
7/10

pH RANGE
6.7–7.4

Rummynose tetra / *Hemigrammus rhodostomus*

ORIGIN
South America

TEMPERATURE RANGE
22–26°C (72–79°F)

COMMUNITY
very good

ADULT SIZE
female 5.5cm (2.2in.)
male 5.5cm (2.2in.)

DIET
all foods

EASE OF KEEPING
9/10

pH RANGE
6.8–7.5

Another very popular fish that is extremely easy to keep. Rummynose tetras like plenty of fine-leafed plants in the aquarium and react well when kept in shoals of 10 or more, nearly always shoaling together. Rummynose tetras can be added to an aquarium after they have been running for between six and eight weeks, when the bacteria levels in the filtration system are adequate. The most striking feature of this fish is its bright-red nose, from which it gets its name. Long, slender fish, rummynose tetras are fast swimmers. The females tend to have much fuller bodies than the males. They can breed in an aquarium given the right conditions, but as soon as the eggs are scattered by the female and fertilised by the male, the parents try to find the eggs and eat them.

Feeding

Rummynose tetras like a varied diet of different flake foods. They readily take live food such as bloodworm, tubifex or glassworm. It is also advisable to feed live daphnia once or twice a week.

Lemon tetra / *Hyphessobrycon pulchiprinnis*

The lemon tetra is a very peaceful and attractive small fish, with bright-yellow-and-black markings in the dorsal and anal fins. As with most tetras, they do best in groups of five or more. Strong and hardy fish for their size, the lemon tetras' fins are nearly always fully erect when they swim, but if they are not, something may be wrong in the aquarium. They like the security of plenty of tall covering plants, such as giant straight vallis, *Wisteria* or *Cabomba*. They are quite an easy fish to breed, with the fry growing to around 2 cm (1in.) in about eight weeks. Sexing the lemon tetra is very easy, as the females have much thicker bodies than the males.

Feeding

They will take all types of food offered, but do much better on a wide variety of dried, frozen and live foods. Make sure that their food is broken down small enough for them to eat.

ORIGIN
South America

TEMPERATURE RANGE
22–26°C (72–79°F)

COMMUNITY
very good

ADULT SIZE
female 5cm (2in.)
male 5cm (2in.)

DIET
all foods

EASE OF KEEPING
10/10

pH RANGE
6.8–7.5

Black widow tetra / *Gymnocorymbus ternetzi*

ORIGIN
South America

TEMPERATURE RANGE
22–26°C (72–79°F)

COMMUNITY
very good

ADULT SIZE
female 5.5cm (2.2in.)
male 5.5cm (2.2in.)

DIET
all foods

EASE OF KEEPING
10/10

pH RANGE
6.8–7.5

With the two jet-black bars on a disc-shaped, silver body, this fish is exceptionally attractive. The black coloration will fade very quickly when the fish is unsettled, but does return to normal within a few hours. Coloration is always stronger in young fish and fades gently as the fish gets older. The black widow has been known for many years and is an old favourite, swimming proudly around an aquarium with its fins held erect. Breeding can be achieved very easily, but a separate tank is necessary as, like most tetras, black widows are avid egg-eaters. Females are quite easily to distinguish from the males as they are slightly more elongated and, when full of roe, quite full in the body.

Feeding

They will take all types of food, but do much better on a wide variety of dried and frozen. Tubifex, bloodworm and glassworm will also be readily taken and need to be fed on a regular basis. Daphnia once or twice a week will also be good for them.

Silvertip tetra / *Hasemania nana*

Male silvertips have slightly better colouring than the female, with a bronze hue, a black line at the rear of the body and bright-white tips to the fins. With an elongated body, silvertips are fast swimmers and look particularly attractive in small shoals. They do better in a tank full of bushy plants, with a clear area in the central front area of the aquarium. They are quite a peaceful fish, but are very inquisitive when there is a new addition to the aquarium and will probably harmlessly harass it.

Feeding

The silvertip will take all types of food, but favour meaty foods, such as bloodworm and tubifex. They will even eat very small slivers of grated frozen beef heart. They are always among the first to reach food, so make sure that you put enough food for the bottom-feeders in the aquarium. Be careful not to overfeed.

ORIGIN
South America

TEMPERATURE RANGE
22–26°C (72–79°F)

COMMUNITY
very good

ADULT SIZE
female 5cm (2in.)
male 5cm (2in.)

DIET
all foods

EASE OF KEEPING
10/10

pH RANGE
6.8–7.5

Glowlight tetra / *Hemigrammus erythrozonus*

ORIGIN
South America

TEMPERATURE RANGE
22–26°C (72–79°F)

COMMUNITY
very good

ADULT SIZE
female 4.5cm (1.75in.)
male 4.5cm (1.75in.)

DIET
all foods

EASE OF KEEPING
9/10

pH RANGE
6.8–7.5

The glowlight tetra is an extremely peaceful fish that is very easy to keep. The body is mainly semi-translucent, with a bright-reddish-gold line through the body length; the dorsal fin has a small hint of red and the anal and pelvic fins have white tips, with the remainder of the fins being clear. They like to be kept, as with most tetras, in small shoals and with other small fish. The females are fuller in the body than the males. Given the right conditions, they can live for four to five years. Breeding is quite difficult, as the water conditions have to be correct, but if you succeed, the fry are easy to raise and grow quite quickly.

Feeding

These fish will take all types of food, but need a good variety of quality foods. Live foods are readily accepted, but need to be part of a varied diet. Make sure that whatever you feed is small enough.

Flame tetra / *Hyphessobrycon flammeus*

This fish likes plenty of plant cover and places to hide. It will take up to a week for the flame tetra to adjust to a new aquarium, but, once settled, is a pleasure to watch and easy to keep. The males tend to look smaller than the females because the females fill out more. The males' colour is much stronger, with deep-red coloration in the body and fins and black edges to the fins. They are easy to tell apart. The flame tetra is also very easy to breed and is ideal for the novice aquarist.

Feeding

The flame tetra will take most types of food, but needs a good variety that is small enough for them to consume. Live foods are readily accepted.

ORIGIN
South America

TEMPERATURE RANGE
22–26°C (72–79°F)

COMMUNITY
very good

ADULT SIZE
female 4.5cm (1.75in.)
male 4.5cm (1.75in.)

DIET
all foods

EASE OF KEEPING
10/10

pH RANGE
6.8–7.5

Black neon tetra / *Hyphessobrycon herbertaxelrodi*

ORIGIN
South America

TEMPERATURE RANGE
22–26°C (72–79°F)

COMMUNITY
very good

ADULT SIZE
female 4.5cm (1.75in.)
male 4.5cm (1.75in.)

DIET
all foods

EASE OF KEEPING
10/10

pH RANGE
6.8–7.5

The black neon tetra is a bold and hardy fish, with both the male and female being quite thick-bodied. It is a very peaceful fish, although it does look a little pugnacious. Body markings consist of a black bar running the full length of the body and a gold-to-faint-green line above this, depending upon the light. The fins are mostly clear, but there is a milky-white colour around the edges of the dorsal and tail fin. It is a very attractive fish. Black neon tetras swim mid-water and will mix quite readily with all other small fish. They are easy to sex, as the female is fuller in the body, and they are also very easy to breed. They are recommended to the novice fish-breeder.

Feeding

Black neon tetras will take most types of food and need a good variety of quality foods. Live foods are readily accepted, particularly meaty types, such as tubifex or bloodworm.

Bleeding-heart tetra / *Hyphessobrycon erythrostigma*

Most bleeding-heart tetras originate from the seemingly unlimited quantities caught in the wild, mainly in Peru. This fish prefers slightly softer water than most tetras and will look its best in this. It is an exceptionally attractive fish, with a deep-red hue all over the body, a dark-red spot in the centre, a long, flowing dorsal fin on the male and a rounded dorsal fin on the female. The dorsal fin on the male is mainly black, turning to red at its base, whereas the female's has a splash of colour at the top, which is pink, turning to bright white. They look really good in small shoals against bright-green aquarium plants. The difference in coloration makes them very easy to sex. Breeding is possible, but not easy.

Feeding

This fish will take most types of food, but need a good variety of quality foods. Live foods are readily accepted, particularly live bloodworm.

ORIGIN
South America

TEMPERATURE RANGE
22–26°C (72–79°F)

COMMUNITY
excellent

ADULT SIZE
female 7cm (2.75in.)
male 7cm (2.75in.)

DIET
all foods

EASE OF KEEPING
9/10

pH RANGE
6.6–7.2

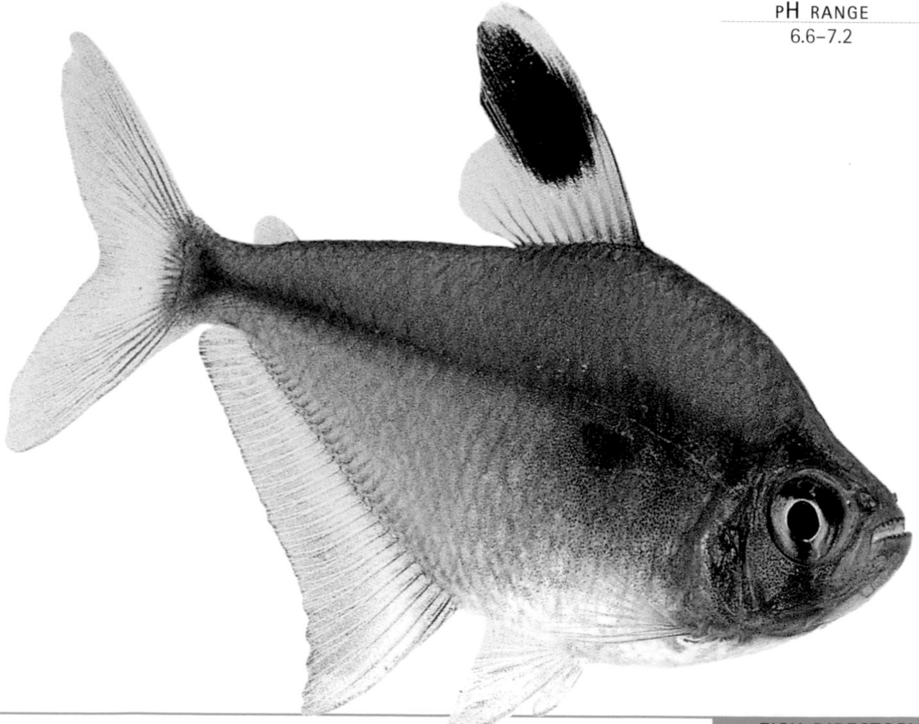

Serpae tetra / *Hyphessobrycon serpae*

ORIGIN
South America

TEMPERATURE RANGE
22–26°C (72–79°F)

COMMUNITY
very good

ADULT SIZE
female 4.5cm (1.75in.)
male 4.5cm (1.75in.)

DIET
all foods

EASE OF KEEPING
10/10

pH RANGE
6.8–7.5

Serpae tetras are exceptionally hardy and resilient fish; four or five are ideal to place in an aquarium when it is first set up. With a red body, a black dorsal fin with white edges and a black blotch behind the eye, it is an attractive fish. The female is very easily distinguished from the male by the fuller body when in breeding condition. They chase other fish in the aquarium, but do no damage. The serpae tetra is one of the easier tetras to breed, but must be set up on its own for this to be successful. They are excellent fish for the novice aquarist.

Feeding

Serpae tetra will take all types of food, but to keep them in top condition they need a good variety. Live foods are readily accepted.

Purple emperor tetra / *Inpaichthys kerri*

This fish is also known as the blue emperor tetra because of the different shades of coloration. It is a beautiful fish, originating in a wide area of the river Amazon. Sexing is very simple: both sexes have the same base marking of a wide stripe running the full length of the body, but only the males have a purple or blue coloration. When first placed in an aquarium, the fish's colours will look washed out, but they will return after a day or two of settling in. Maintained correctly, these fish grow to about 5cm (2in.) and have a thick, deep body. As with most tetras, they will benefit from the security of a well-planted tank.

Feeding

Purple emperor tetras need a good variety of quality foods. Glassworm and daphnia will be chased around the tank until it has all been eaten, and other live foods are also readily accepted.

ORIGIN
South America

TEMPERATURE RANGE
22–26°C (72–79°F)

COMMUNITY
very good

ADULT SIZE
female 5cm (2in.)
male 5cm (2in.)

DIET
all foods

EASE OF KEEPING
8/10

pH RANGE
6.8–7.5

Black emperor tetra / *Nematobrycon palmeri* (black)

ORIGIN
South America

TEMPERATURE RANGE
22–26°C (72–79°F)

COMMUNITY
very good

ADULT SIZE
female 5cm (2in.)
male 5cm (2in.)

DIET
all foods

EASE OF KEEPING
8/10

pH RANGE
6.8–7.5

This fish is a close relative of the purple emperor tetra (see page 45) and is a colour variety of the standard emperor tetra (see page 47). In a good specimen, the black coloration covers the entire body and the eye has a metallic-blue ring around it. Once adult, the males grow extensions to the top and bottom rays of the tail and a third extension in the centre of the other two. If the tank conditions are correct, they are very easy to breed. They need plenty of tall, fine-leafed, feathery plants (e.g., *Cabomba, Nitella*) and spawn among them, leaving the eggs attached to the leaves of the plants.

Feeding

Black emperor tetras need a good variety of quality foods, including live foods such as glassworm and daphnia.

Emperor tetra / *Nematobrycon palmeri*

This is generally considered to be the original emperor tetra. Found many years ago, it has always been a favourite of the aquarium-keeper. With a bright-blue ring to the eye, a black bar running the length of the body and metallic-blue-and-red coloration in the upper half of the body, it is a beautiful addition to any aquarium. Although this fish prefers slightly acidic water with less hardness, it does extremely well once it has become accustomed to the local water conditions. Because it is naturally a shoaling fish, it is advisable to keep a number of these fish together in your aquarium. The male emperor tetra also has three extensions to the caudal fin.

Feeding

Emperor tetras need a good variety of quality foods, including live feed, such as glassworm and daphnia.

ORIGIN
South America

TEMPERATURE RANGE
22–26°C (72–79°F)

COMMUNITY
very good

ADULT SIZE
female 5cm (2in.)
male 5cm (2in.)

DIET
all foods

EASE OF KEEPING
8/10

pH RANGE
6.8–7.5

Black phantom tetra / *Megalamphodus megalopterus*

ORIGIN
South America

TEMPERATURE RANGE
22–26°C (72–79°F)

COMMUNITY
very good

ADULT SIZE
female 4cm (1.5in.)
male 4cm (1.5in.)

DIET
all foods

EASE OF KEEPING
8/10

PH RANGE
6.8–7.5

Originating in Colombia, in South America, this fish is a relative newcomer to aquariums. When first discovered about 35 years ago, it was regarded as a very rare and difficult fish to keep and breed, but this myth was quickly dispelled. With a striking black hue to the body and fins, and jet-black markings on the dorsal fin when settled and in good condition, the black phantom is very popular among hobbyists and looks very effective in a shoal of 10 to 15. The female has a fuller body than the male when she is full of roe. Breeding is relatively easy, and the young grow on well. Some excellent examples are commercially bred in the Czech Republic that are very strong in size, colour and body.

Feeding

This fish likes its food to be on the small side. Flake food needs to be rubbed between your fingers to break it down. Small, live food is also readily taken, such as baby daphnia, microworms and small bloodworms.

Red phantom tetra / *Megalamphodus sweglesi*

A close relative of the black phantom tetra (see page 50), this is also a recent addition to European aquariums. Discovered shortly after the black phantom, it was also regarded as a sensitive fish. It has a striking red hue to the body and fins, black markings on the dorsal fin and a small, bright-white splash on the dorsal fin of the female. When you can get this fish to breed, the young are very easy to grow on, and successful commercial breeding is carried out in the Czech Republic. There are two other phantoms in this genus, but they are rarely seen and are normally only imported by accident.

Feeding

This fish likes small food. Flake food must to be rubbed between your fingers to break it down. Small live food is also readily taken, such as baby daphnia, microworms and small bloodworms.

ORIGIN
South America

TEMPERATURE RANGE
22–26°C (72–79°F)

COMMUNITY
very good

ADULT SIZE
female 4cm (1.5in.)
male 4cm (1.5in.)

DIET
all foods

EASE OF KEEPING
7/10

pH RANGE
6.8–7.5

Red-eye tetra / *Moenkhausia sanctaefilomenae*

ORIGIN
South America

TEMPERATURE RANGE
22–26°C (72–79°F)

COMMUNITY
good

ADULT SIZE
female 10cm (4in.)
male 10cm (4in.)

DIET
all foods

EASE OF KEEPING
10/10

pH RANGE
6.8–7.8

A very tough and hardy fish, the red-eye tetra is ideal to place in a new aquarium. Both male and female are thick-bodied fish and can be quite boisterous, but if kept in a small shoal of five or six they will chase each other and leave other species alone. The body is silver, with a black spot at the root of the tail and a half-circle of red above the eye. If this fish is unhappy or unwell, the coloration will change to silvery-black. It is very easy to breed, but the adults must be removed as soon as they have finished spawning.

Feeding

Red-eye tetras will take just about anything that is offered, including grated beef heart, chopped small earthworm, grindleworm and so on. They will continue to eat while there is food in the aquarium, so it is advisable to feed only small quantities at a time, ensuring that there is enough food for the other occupants of the aquarium.

Three-lined pencilfish / *Nannostomus trifasciatus*

This fish does exceptionally well when in a species tank or a shoal of 15 or more. They swim pointing slightly nose downwards and are a sedate fish, troubling none of their fellow occupants. With a golden body, black lines from nose to tail and bright-red splashes on the fins and body, they are attractive additions to an aquarium. When breeding, these fish are egg-scatterers and like to spawn in plants such as Java moss, where the eggs can drop between the fine threads and be hidden until they hatch. When the young are free-swimming, they require foods like microsorium, infusoria and newly hatched artemia. They will also use this type of plant to hide in when they feel threatened

Feeding

This fish has a very small, tubular mouth and needs small food.
Flake food must be rubbed between your fingers to break it down, and small, live food is also readily taken, such as baby daphnia, microworms, brine shrimps and small bloodworms.

ORIGIN
South America

TEMPERATURE RANGE
22–26°C (72–79°F)

COMMUNITY
excellent

ADULT SIZE
female 5cm (2in.)
male 5cm (2in.)

DIET
all foods

EASE OF KEEPING
9/10

pH RANGE
6.8–7.5

Marginatus pencilfish / *Nannostomus marginatus*

ORIGIN
South America

TEMPERATURE RANGE
22–26°C (72–79°F)

COMMUNITY
excellent

ADULT SIZE
female 3.5cm (1.4in.)
male 3.5cm (1.4in.)

DIET
all foods

EASE OF KEEPING
8/10

pH RANGE
6.8–7.5

Originating in Surinam and Guyana, marginatus pencilfish are found in prolific numbers, breeding very freely in their natural habitat. It does exceptionally well when in a species tank or a shoal of 15 or more and is a very sedate fish. It can be mistaken for the three-lined pencilfish (see page 53), except that the pelvic and anal fins are bright red. When these fish are ready to breed, they scatter the eggs into very fine plants, where they are left to hatch without any parental care. When the young are free-swimming, they require foods like microsorium, infusoria and newly hatched artemia and hide in this type of plant when they feel threatened.

Feeding

As with most other pencilfish, they like their food to be on the small side. Flake food needs to be rubbed between your fingers to break it down. Small, live food is also readily taken, such as baby daphnia, microworms and small bloodworms.

Harrisonii pencilfish / *Nannostomus harrisonii*

The harrisonii pencilfish is one of the largest in its group. It has a black, mottled bar running through its eye to the very end of its tail, and there is a splash of red above the bar in the tail. The base body colour is olive. It has a long, tubular-shaped body and mostly swims nose up in the aquarium. The female can be recognised by the slightly fuller body when in spawning condition. All of the *Nannostomus* species of pencilfish are community fish and are very easy to keep. These fish will also mix together without any problems and like plenty of plants in which to swim.

Feeding

They like small food. Flake food must be rubbed between your fingers to break it down, and small, live food is also readily taken, such as baby daphnia, microworms and small bloodworms.

ORIGIN
South America

TEMPERATURE RANGE
22–26°C (72–79°F)

COMMUNITY
excellent

ADULT SIZE
female 6cm (2.4in.)
male 6cm (2.4in.)

DIET
all foods

EASE OF KEEPING
8/10

pH RANGE
6.8–7.5

Congo tetra / *Phennocogrammus interuptus*

ORIGIN
Africa

TEMPERATURE RANGE
22–26°C (72–79°F)

COMMUNITY
excellent

ADULT SIZE
female 9cm (3.5in.)
male 9cm (3.5in.)

DIET
all foods

EASE OF KEEPING
9/10

pH RANGE
6.8–7.5

The male Congo tetra is a really deep, thick-bodied fish. The fins are elongated into filaments and adults have a central extension in the tail. The coloration in this fish lightens or intensifies depending on how light strikes it. There is a wide band of greenish blue that blends into the upper- and lower-body colours, a light-gold bar in the upper body, and the lower body is silvery. None of these colours have definitive edges. The dorsal fin is pointed and there is also a white edging to the fins. The female lacks the tail extension and the dorsal fin is rounded. They can grow to as much as 9cm (3.5in.) in length and need plenty of space in which to swim, so a good-sized aquarium is necessary. They are good community fish and do not harass smaller fish.

Congo tetra, male.

Congo tetra, female.

Feeding

This fish will take almost anything that you offer, from flake food to chopped earthworm.

Red Congo tetra / *Bathyaethiops breuseghemi*

ORIGIN
Africa

TEMPERATURE RANGE
22–26°C (72–79°F)

COMMUNITY
very good

ADULT SIZE
female 3.5cm (1.4in.)
male 3.5cm (1.4in.)

DIET
all foods

EASE OF KEEPING
8/10

pH RANGE
6.8–7.5

There are five or six species in this group that share the same common name. This particular species is the most common, but has only recently been acquired by aquarists. When the fish is young you can see the potential of the coloration within it, but it is not until it has gained maturity that it shows its full splendour. The body shape is oval, with a base colour of light olive green, a small, dark spot behind the gill cover, a large, dark spot at the end of the body and a bright-red mark in the dorsal fin, mainly towards the base of the leading rays. The body colour changes as the fish turns in the light.

Feeding

This fish much prefers live foods, such as bloodworms or tubifex. After a week or so of settling into the aquarium it will take frozen foods without any problems. Large, flake food is also readily accepted.

Bloodfin tetra / *Aphyocharax annisitsi*

Three fish share the common name of bloodfin tetra, but when seen together they are clearly very different. *Aphyocharax annisitsi* is the most common. It is a slim, elongated fish, and the rear, lower area of the body seems to have been severely pinched. This is its natural shape. The body is semi-translucent, with areas of bright red in the lower half and in the tail. It will take two or three days for this fish to settle down and gain its full colour. It will be quite happy in ones, twos or a small shoal.

Feeding

The bloodfin tetra will readily take most foods, but can be rather slower than the other occupants when you feed live foods. After a week or so of settling into the aquarium they will take frozen foods without any problems. Large, flake food is also readily accepted.

ORIGIN
South America

TEMPERATURE RANGE
22–26°C (72–79°F)

COMMUNITY
very good

ADULT SIZE
female 6cm (2.4in.)
male 6cm (2.4in.)

DIET
all foods

EASE OF KEEPING
9/10

pH RANGE
6.8–7.5

Rosy tetra / *Hyphessobrycon roseaceus*

ORIGIN
South America

TEMPERATURE RANGE
22–26°C (72–79°F)

COMMUNITY
excellent

ADULT SIZE
female 4.5cm (1.7in.)
male 4.5cm (1.7in.)

DIET
all foods

EASE OF KEEPING
9/10

pH RANGE
6.8–7.5

When fully settled in the aquarium, this is one of the prettiest tetras available. The body has an all-over red hue and a more solid red in the anal and caudal fins. The male's dorsal fin is quite long and comes to a point, whereas the female's is much shorter and rounder, with a pinkish-red edging to the upper part. The rosy tetra normally swims with its fins fully extended, making a very showy fish. It prefers fish of a similar size as fellow tank mates. It flourishes in softer water, but will accept normal tap water. Breeding is possible, but quite difficult unless conditions are correct.

Feeding

This fish prefers live foods, such as bloodworms or tubifex. After a week or so of settling into the aquarium they will take frozen foods without any problems. Flake food is also readily accepted, but will have to be broken down between your fingers.

X-ray tetra / *Pristella maxillaris*

This fish does not look quite as its name suggests. The swim bladder and scale line markings are visible, but the rest of the body is opaque. The caudal and anal fins are red, and the dorsal fin has black, yellow and white coloration, so it is a very pretty fish. It is also an extremely hardy fish, adapting to most water conditions. It is a community fish and is one of the stronger fish that could be placed in a new aquarium. They will normally swim together if you place a shoal of them in the tank and prefer a well-planted aquarium. Extremely easy to keep, they are one of the easiest tetras to breed, but you must remove the adults once they have spawned.

Feeding

These fish will eat all types of live, frozen and flake food. They are exceptionally fast to feed when food is placed in the tank, to the detriment of other fish. Beware of overfeeding.

ORIGIN
South America

TEMPERATURE RANGE
22–26°C (72–79°F)

COMMUNITY
excellent

ADULT SIZE
female 4.5cm (1.7in.)
male 4.5cm (1.7in.)

DIET
all foods

EASE OF KEEPING
10/10

pH RANGE
6.8–7.5

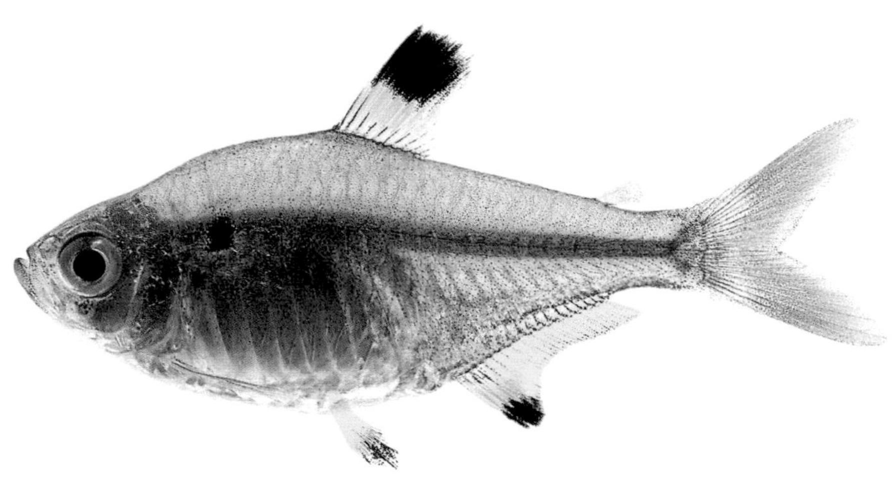

Penguin tetra / *Thayeria boehlkei*

ORIGIN
South America

TEMPERATURE RANGE
22–26°C (72–79°F)

COMMUNITY
very good

ADULT SIZE
female 6.5cm (2.5in.)
male 6.5cm (2.5in.)

DIET
all foods

EASE OF KEEPING
10/10

pH RANGE
6.8–7.5

This is a very striking fish, with a thick, black band running the full length of the body through to the lower lobe of the tail, a silvery lower half of the body and a shiny, golden upper half. When stationary, the fish slowly drops its tail to a 45° position and then, with a flick of the tail, brings the body back up into line again. Penguin tetras will chase other fish from time to time, but no damage is done. Females have a much deeper belly, but both sexes are rather stocky fish. Breeding the penguin tetra can be quite difficult, although if you are successful the youngsters are easy to grow on. They do very well in small shoals of between five and ten fish.

Feeding

These fish will eat all types of live, frozen and flake food. They are exceptionally fast to feed when food is placed in the tank, to the detriment of other fish. Beware of overfeeding.

Beacon tetra / *Hemigrammus occelifer*

The beacon tetra is also sometimes known as the 'head-and-tail-light tetra', from the bright-red colour in the upper half of the eye and the golden spot in the upper caudal peduncle. There is a black line running from just below the dorsal fin to the caudal, ending in the gold spot. The body is semi-translucent, with the fins being clear except for the pectoral and anal fins, which have white in the front rays. Both males and females are deep bodied, but when seen from the top it is clear that the male is the slimmer fish. Beacon tetras enjoy having plenty of plants in the aquarium. They are an ideal fish for novice aquarists because they are so easy to keep and to breed.

Feeding

These fish will readily accept all types of food. They are not an exceptionally fast fish to feed, so ensure that there is plenty of food for them without overfeeding.

ORIGIN
South America

TEMPERATURE RANGE
22–26°C (72–79°F)

COMMUNITY
excellent

ADULT SIZE
female 5cm (2in.)
male 5cm (2in.)

DIET
all foods

EASE OF KEEPING
10/10

pH RANGE
6.8–7.5

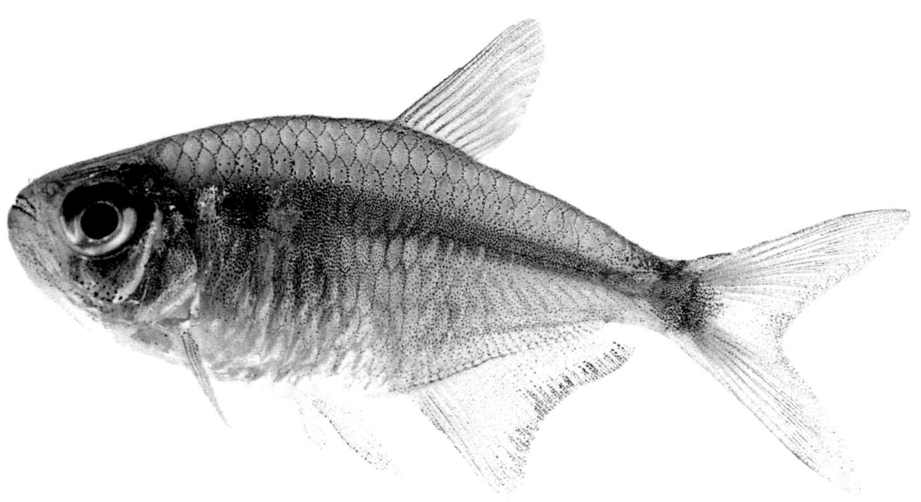

Loretto tetra / *Hyphessobrycon loretoensis*

ORIGIN
South America

TEMPERATURE RANGE
22–26°C (72–79°F)

COMMUNITY
excellent

ADULT SIZE
female 4.5cm (1.7in.)
male 4.5cm (1.7in.)

DIET
all foods

EASE OF KEEPING
8/10

pH RANGE
6.8–7.5

When first imported from the wild, these fish are very skinny and lack colour, which implies that there is not a great deal of food available in their natural habitat. It will take about 14 days for the body to fill out if you give three to four small feeds a day. The colour will also intensify, revealing a truly beautiful fish. They have a strong red in their white-tipped fins, the upper half of the body is gold and a wide, black band runs the length of the body. They do exceptionally well in a shoal, especially in a well-planted aquarium. Loretto tetras prefer to be with fish of their own size.

Feeding

These fish will readily accept most types of food. High-quality food, such as brine shrimps or bloodworms, is vital when first placing them in your aquarium. Once the colour and body shape is established, a more varied diet should be offered.

Blind cave fish / *Astyanax fasciatus mexicanus*

Blind cave fish are naturally blind because they are found in subterranean caves where sight is of no use to them. They have naturally evolved a 'radar system' so that they rarely bump into other fish or items in your aquarium. They are nearly always the first to sense that food has been placed in the aquarium and have no problems finding it. With a thickset body and a cream-to-pinkish body coloration, they will certainly stand out in an aquarium. They swim in a very docile manner until food is placed in the aquarium, when they become unexpectedly fast. There are no problems at all with other fish, and they are quite easy to breed.

Feeding

You will have no problems getting these fish to feed. They will readily accept all types of food.

ORIGIN
Central America

TEMPERATURE RANGE
22–26°C (72–79°F)

COMMUNITY
very good

ADULT SIZE
female 10cm (4in.)
male 10cm (4in.)

DIET
all foods

EASE OF KEEPING
10/10

pH RANGE
6.8–7.5

Splashing tetra / *Pyrrhulina sp.*

ORIGIN
South America

TEMPERATURE RANGE
22–26°C (72–79°F)

COMMUNITY
excellent

ADULT SIZE
female 7cm (2.75in.)
male 7cm (2.75in.)

DIET
all foods

EASE OF KEEPING
8/10

pH RANGE
6.8–7.5

Once they have settled into an aquarium, which will take a week or two, these fish are easy to keep. Splashing tetras require plenty of bushy, fine-leafed plants in the aquarium. They can escape by jumping through any small gaps in the top of an aquarium, so ensure that there are no holes. In their natural habitat they breed by actually jumping out of the water and adhering their eggs to the underside of overhanging plant leaves. They can be induced to spawn in an aquarium, although when hatched out the fry are minuscule and can be very difficult to rear.

Feeding

These fish will readily accept all types of food. They are not an exceptionally fast fish to feed, so ensure that there is plenty of food for them without overfeeding.

Colombia 95 tetra / *Astyanax sp.*

The Colombia 95 has only been known to the hobby for a short time, hence the rather utilitarian name. The male is by far the more colourful, with a vivid, red tail, adipose fin and anal fin, and as it gets older the whole body becomes a deep blue. As this fish turns under the aquarium lights, you can see a myriad of blues in the body. Although a new fish, they have already been found to breed very easily. A pair of healthy fish can be placed in a tank containing fine-leafed plants on their own. When they start to breed they will lay between six and ten eggs per day. Once spawning, you can leave the adults to continue to spawn without any fear of them eating the eggs. You should rest them every two weeks for a while so that they can be brought back into top condition.

Feeding

These fish will eat all types of live, frozen and flake food. To get the best from this fish, only feed the very best-quality foods available.

ORIGIN
South America

TEMPERATURE RANGE
22–26°C (72–79°F)

COMMUNITY
excellent

ADULT SIZE
female 7cm (2.75in.)
male 7cm (2.75in.)

DIET
all foods

EASE OF KEEPING
10/10

pH RANGE
6.8–7.5

Cherry barb / *Barbus titteya*

ORIGIN
India/Sri Lanka

TEMPERATURE RANGE
22–26°C (72–79°F)

COMMUNITY
excellent

ADULT SIZE
female 5cm (2in.)
male 5cm (2in.)

DIET
all foods

EASE OF KEEPING
10/10

pH RANGE
6.8–7.5

Barbs have a poor reputation as community fish, but the cherry barb is an exception and makes a suitable addition to any aquarium. Only when the fish have grown to maturity does the male show the full colour from which its name is derived. When in breeding condition, the fish is cherry red all over, fins included, but the female has weaker coloration. They are an exceptionally popular fish because of their peaceful nature and attractiveness. When ready to court, the male swims around the female, spreading his fins to such an extent that you would think that they would split. When they have laid their eggs the adults must be removed, otherwise they will go hunting for the eggs and eat them.

Feeding

These fish will eat all types of live, frozen and flake food. When small, ensure that the flake food is broken down.

Tiger barb / *Barbus tetrazona*

The tiger barb gets its name from its markings and coloration. They will chase other fish and nip their fins sometimes, but if kept in a group of seven or eight fish they tend to leave the other inhabitants alone. With a golden body, four black bars running completely around the body and red fins, this is a very attractive fish. The tiger barb is commercially bred and is therefore easy to obtain. They are ideal fish for the novice aquarist to breed, but, as with most barbs, must be taken away from the eggs once they are laid, otherwise they will eat them. Their fry are easy to raise.

Feeding

Tiger barbs will eat all types of live, frozen and flake food. They will also take such foods as finely grated beef heart. Make sure that they are well fed, otherwise they are liable to start chasing the other fish in your aquarium.

ORIGIN
Asia

TEMPERATURE RANGE
22–26°C (72–79°F)

COMMUNITY
good

ADULT SIZE
female 7cm (2.75in.)
male 7cm (2.75in.)

DIET
all foods

EASE OF KEEPING
10/10

pH RANGE
6.8–7.5

Green tiger barb / *Barbus tetrazona*

ORIGIN
Asia

TEMPERATURE RANGE
22–26°C (72–79°F)

COMMUNITY
good

ADULT SIZE
female 7cm (2.75in.)
male 7cm (2.75in.)

DIET
all foods

EASE OF KEEPING
10/10

pH RANGE
6.8–7.5

The green tiger barb is a colour variation on the standard tiger barb (see page 69). There are other varieties, but the standard tiger and green tiger barb are the most popular. These fish do not seem to be quite as pursuing and boisterous as standard tiger barbs, but it is still advisable to keep them in a group of seven or eight fish. With a velvet-green body coloration, this is a very attractive fish. The green tiger barb is also commercially bred and is very easy to obtain. They are ideal fish for the novice aquarist to breed, but must be taken away from the eggs once they are laid. The fry are easy to raise.

Feeding

Tiger barbs will eat all types of live, frozen and flake food. They will also take such foods as finely grated beef heart. Make sure that they are well fed, otherwise they are liable to start chasing the other fish in the aquarium.

Chequer barb / *Barbus oligolepis*

ORIGIN
Indonesia/Sumatra

TEMPERATURE RANGE
22–26°C (72–79°F)

COMMUNITY
excellent

ADULT SIZE
female 4cm (1.5in.)
male 4cm (1.5in.)

DIET
all foods

EASE OF KEEPING
10/10

pH RANGE
6.8–7.5

This fish has been known to aquarists for many years and is a hardy and peaceful fish. When well-conditioned, the male shows black scale-edge markings down the centre of the body, hence its name. The scales are a rich, silver colour, and the red fins have black edging. The female is much plainer, with a slightly fuller body. The chequer barb swims with the fins erect; if the fins are clamped or lowered for some time, it normally means that they are unhappy or ill, so check them carefully. Breeding the chequer barb is quite easy, but, as with most barbs, remove the adults when they have finished spawning.

Feeding

Chequer barbs will eat all types of live, frozen and flake food. Ensure that they are getting enough food without polluting the aquarium, because they are not the fastest to react to feeding time.

Rosy barb / *Barbus conchonius*

ORIGIN
India

TEMPERATURE RANGE
22–26°C (72–79°F)

COMMUNITY
very good

ADULT SIZE
female 8cm (3in.)
male 8cm (3in.)

DIET
all foods

EASE OF KEEPING
10/10

pH RANGE
6.8–7.5

Rosy barbs are large, stocky fish, and normally swim around an aquarium in a slow, lethargic way unless in the company of other males, when they will show off to each other in their best regalia. The male has a gold base colour to the body, with an overlay of red in the bottom two-thirds of the body. The dorsal fin is red and black, the caudal red and the anal fin is clear, except for a black edging. The female is much plainer, with a gold body and a more distinctive black spot close to the caudal peduncle; when in breeding condition, females are much plumper in the body.

Rosy barb, male.

Feeding

Rosy barbs will eat all types of live, frozen and flake food. They are often the first to feed, so ensure that there is enough food for the other fish in the aquarium.

Rosy barb, female.

Long-fin rosy barb / *Barbus conchonius*

ORIGIN
India

TEMPERATURE RANGE
22–26°C (72–79°F)

COMMUNITY
very good

ADULT SIZE
female 8cm (3in.)
male 8cm (3in.)

DIET
all foods

EASE OF KEEPING
10/10

pH RANGE
6.8–7.5

The long-fin rosy barb is a variant of the standard rosy barb (see pages 72 to 73). The male has a gold base colour to the body, with an overlay of red in the bottom two-thirds of the body. The dorsal fin is red and black, the caudal fin red and the anal fin is clear, except for a black edging. Fins on the male are much more extended. The female is much plainer, with a gold body and a more distinctive black spot close to the caudal peduncle; they are much plumper in the body when in breeding condition.

Long-fin rosy barb, female.

Long-fin rosy barb, male.

Feeding

Long-fin rosy barbs will eat just about all types of live, frozen and flake food. They are often the first to feed, so ensure there is enough food for the other fish in the aquarium.

Golden barb / *Barbus schuberti*

ORIGIN
Asia/China

TEMPERATURE RANGE
22–26°C (72–79°F)

COMMUNITY
excellent

ADULT SIZE
female 10cm (4in.)
male 10cm (4in.)

DIET
all foods

EASE OF KEEPING
10/10

pH RANGE
6.8–7.5

The golden barb has been popular with aquarists for many years. It has a beautiful yellow-to-golden body, with red fins, and really stands out in an aquarium. Adults are quite thick-bodied and deep in their coloration. Males generally have a row of black spots running from just behind the gills to the caudal peduncle, and females usually have a few scattered black spots on their body. Breeding can be difficult, as females are sometimes unwilling to shed their eggs, no matter how much attention the male gives her. If breeding is successful, the fry are very small and need to be fed on microsorium as a starter food.

Feeding

Golden barbs will eat all types of live, frozen and flake food. They are often the first to feed, so ensure there is enough food for the other fish in the aquarium.

Five-banded barb / *Barbus pentazona*

Breeding programmes in the Czech Republic have recently made this fish more widely available. As the name suggests, it has five vertical bars, which are very dark green to black in colour. The base colour of the body is a reflective light gold, with a reddish hue across the back, and the dorsal, pelvic and anal fins are red, so it is a distinctive fish. They look exceptional when in small shoals, preferring to swim together in the mid- to lower half of the aquarium. The male is slightly brighter in coloration than the female, and also slightly slimmer. They like plenty of broad-leafed plants in which to take cover if frightened.

Feeding

These barbs will eat all types of live, frozen and flake food. They swim in the lower half of the aquarium and will not be among the first to feed, so ensure that they get enough to eat without overfeeding.

ORIGIN
South-east Asia

TEMPERATURE RANGE
22–26°C (72–79°F)

COMMUNITY
excellent

ADULT SIZE
female 5cm (2in.)
male 5cm (2in.)

DIET
all foods

EASE OF KEEPING
10/10

pH RANGE
6.8–7.5

Stolicks barb / *Barbus stoliczkanus*

ORIGIN
Asia

TEMPERATURE RANGE
22–26°C (72–79°F)

COMMUNITY
excellent

ADULT SIZE
female 5cm (2in.)
male 5cm (2in.)

DIET
all foods

EASE OF KEEPING
10/10

pH RANGE
6.8–7.5

Stolicks barb closely resembles ticto barb, and there is some dispute over whether they are separate species. Whatever the answer, this is a very peaceful and beautiful addition to any aquarium. The base body colour is bronze, with two black spots in both male and female; the male has a wide, red-bar overlay of colour extending the full length of the body. The dorsal fin on the male also has red and black flecking within it. These fish are not widely available, so are worth purchasing if you see them. To breed, put one male with several females and remove the eggs as soon as they have finished spawning.

Feeding

These barbs will eat all types of live, frozen and flake food. They swim in the lower half of the aquarium and will not be among the first to feed, so ensure that they get enough to eat without overfeeding.

Cape lopez lyretail (gold) / *Aphyosemion australe*

For many years killifish have been regarded as a specialist fish, but, with a few exceptions, the species now available are community fish. The Cape Lopez lyretail, also very well known by its Latin name, is one of the most widely available. With bright reds, yellows, metallic blues and greens in the body and fin coloration, and white top-and-bottom extensions to the tail, this is a beautiful fish. Their average life span in a community tank is about two years, but they are very easy to breed.

Feeding

Killifish will eat all types of live, frozen and flake food. They swim in the lower half of the aquarium and will not be among the first to feed, so ensure that they get enough to eat without overfeeding.

ORIGIN
West Africa

TEMPERATURE RANGE
22–26°C (72–79°F)

COMMUNITY
excellent

ADULT SIZE
female 5cm (2in.)
male 6cm (2.4in.)

DIET
all foods

EASE OF KEEPING
10/10

pH RANGE
6.8–7.5

Steel-blue aphyosemion (gold) / *Aphyosemion gardneri*

ORIGIN
West Africa

TEMPERATURE RANGE
22–26°C (72–79°F)

COMMUNITY
excellent

ADULT SIZE
female 5cm (2in.)
male 6cm (2.4in.)

DIET
all foods

EASE OF KEEPING
10/10

pH RANGE
6.8–7.5

With red spotting over a golden body, and bright red and yellow in the tail, this is a very colourful, community fish, although the females in the genus *Aphyosemion* are all very plain and drab. Killifish can be divided into two groups: mop-spawners and peat-divers. *Aphyosemion* are mop-spawners and the fish attach their eggs to lengths of wool suspended in the water from a cork. The spawn are then removed from the aquarium and placed in a warm area for a short period, before being transferredto a small tank where the eggs will hatch out and the fry can be grown on.

Aphyosemion gardneri gold, male.

Aphyosemion gardneri gold, female.

Feeding

Killifish will eat all types of live, frozen and flake food. They swim in the lower half of the aquarium and will not be among the first to feed, so ensure that they get enough to eat without overfeeding.

Orange-fringed killie / *Aphyosemion scheeli*

ORIGIN
West Africa

TEMPERATURE RANGE
22–26°C (72–79°F)

COMMUNITY
excellent

ADULT SIZE
female 5cm (2in.)
male 6cm (2.4in.)

DIET
all foods

EASE OF KEEPING
10/10

pH RANGE
6.8–7.5

This killifish is a blaze of colour, with a metallic-blue body colour, bright-red spots all over it and light yellow, dark yellow and bright red in the fins. It is difficult to understand why we do not see more of these fish when they have this variety of colorations and markings. It is very difficult to identify the markings of individual species without either a specialist book on killifish or a knowledgeable vendor.

Feeding

Killifish will eat all types of live, frozen and flake food. They swim in the lower half of the aquarium and will not be among the first to feed, so ensure that they get enough to eat without overfeeding.

Red-striped killie / *Aphyosemion striatum*

With lines of bright-red, small spots joined together along the full length of its body over a reflective-green body colour, this is another striking fish. The dorsal fin has two thick, bright-red lines and the tail is covered in a red, yellow, orange and blue coloration. Females are much less colourful. Like most killies, they enjoy being in an aquarium that has plenty of fine-leafed plants to hide among or spawn on. It is a good community fish that will be an asset to your aquarium.

Feeding

Killifish will eat all types of live, frozen and flake food. Because they swim in the lower half of the aquarium they will not be among the first to feed, so ensure that they get enough to eat without overfeeding.

ORIGIN
West Africa

TEMPERATURE RANGE
22–26°C (72–79°F)

COMMUNITY
excellent

ADULT SIZE
female 5cm (2in.)
male 6cm (2.4in.)

DIET
all foods

EASE OF KEEPING
10/10

pH RANGE
6.8–7.5

Rachows killie / *Nothobranchius rachovii*

ORIGIN
Africa

TEMPERATURE RANGE
22–26°C (72–79°F)

COMMUNITY
excellent

ADULT SIZE
female 5cm (2in.)
male 6cm (2.4in.)

DIET
all foods

EASE OF KEEPING
10/10

pH RANGE
6.8–7.5

Nothobranchius killifish originate in Africa, and Rachows killie comes from Mozambique. Like most killifish, they do not like a large change in water conditions, so check the pH of the water that they are being kept in and gradually adjust them to new conditions. Although they prefer slightly soft, brown-peat water, they will accept harder tap water after a period of adjustment. They are peat-divers, meaning that the female dives into a peat base to lay her eggs. The male follows directly after her to fertilise them. The peat is taken from the aquarium and dried out. It is then, at a later date, replaced into a small tank containing about 5cm (2in.) of water to hatch the eggs.

Feeding

Killifish will eat all types of live, frozen and flake food. Because they swim in the lower half of the aquarium they will not be among the first to feed, so ensure that they get enough to eat without overfeeding.

Zebra danio / *Brachydanio rerio*

Zebra danios are both one of the hardiest fish available and extremely easy to keep. They are ideal for a new tank because of their adaptability; they are peaceful and fit in well in community tanks and are perfect for the first-time fish breeder because they are so prolific. Both males and females have gold and blue stripes running the length of the body and tail, but the female becomes slightly plumper in the body. The body shape is long and slim, enabling them to swim extremely fast. When attempting to catch them, use as large a net as possible because of their speed. Breeding is extremely easy, but the fry are minuscule, and many batches have been thrown away because they are so difficult to see for the first four or five days after hatching.

Feeding

Zebra danios will eat all types of live, frozen and flake food. They swim at all levels of the aquarium and, no matter where they are, will always be one of the first to get to the food at feeding time.

ORIGIN
India

TEMPERATURE RANGE
22–26°C (72–79°F)

COMMUNITY
excellent

ADULT SIZE
female 5cm (2in.)
male 5cm (2in.)

DIET
all foods

EASE OF KEEPING
10/10

pH RANGE
6.8–7.5

Leopard danio / *Brachydanio frankei*

ORIGIN
South America

TEMPERATURE RANGE
22–26°C (72–79°F)

COMMUNITY
excellent

ADULT SIZE
female 5cm (2in.)
male 5cm (2in.)

DIET
all foods

EASE OF KEEPING
10/10

pH RANGE
6.8–7.5

Leopard danios are hardy, easy to keep and prolific breeders, so they are ideal for new aquariums, community tanks and first-time fish breeders. Both sexes have a beautiful golden body, with very dark-blue spots all over it, and the the female is slightly plumper in the body. With a long, slim body, they are extremely fast swimmers. Breeding is very easy, but remember that the fry are tiny and will be difficult to see for the first four or five days after hatching.

Feeding

As with all danios, these fish will eat all types of live, frozen and flake food. They swim at all levels of the aquarium and, no matter where they are, will always be one of the first to get to the food at feeding time.

Pearl danio / *Brachydanio albolineatus*

This is another extremely hardy and adaptable fish for the new aquarium and novice aquarist. Like all danios, it is a shoaling fish and can easily be kept in quite large shoals of 20 or more without any problems. Both sexes have a steel-blue body, with a very thin, bright-red line running the length of the body. There is also a reddish hue along the ventral area of the male. The female is a much fuller-bodied fish than the male. Breeding is very easy. Place twice as many males as females in the breeding tank and remove the adults as soon as spawning is finished, otherwise they will hunt for, and devour, the eggs.

Feeding

Pearl danios will eat all types of live, frozen and flake food. They swim at all levels of the aquarium and, no matter where they are, will always be one of the first to get to the food at feeding time.

ORIGIN
Asia

TEMPERATURE RANGE
22–26°C (72–79°F)

COMMUNITY
excellent

ADULT SIZE
female 5cm (2in.)
male 5cm (2in.)

DIET
all foods

EASE OF KEEPING
10/10

pH RANGE
6.8–7.5

Giant danio / *Danio aequipinnatus*

ORIGIN
Asia

TEMPERATURE RANGE
22–26°C (72–79°F)

COMMUNITY
very good

ADULT SIZE
female 10cm (4in.)
male 10cm (4in.)

DIET
all foods

EASE OF KEEPING
10/10

pH RANGE
6.8–7.5

The giant danio is one of the larger danios and grows to about 10cm (4in.) in length, with a deep body. It is a very lively, fast-swimming, boisterous fish, with no bad characteristics. Both male and female are very pretty, with a mottled, steel-blue, irregular pattern on a golden body base colour. As the fish swim and turn under aquarium lighting, the blue changes to many different shades, all very attractive. When in breeding condition, the female is slightly deeper and fuller in the belly area. As with all danios, they are easy to breed and, once the fry are feeding, are very easy to rear.

Feeding

Giant danios will eat all types of live, frozen and flake food. They swim at all levels of the aquarium and, no matter where they are, will always be one of the first to get to the food at feeding time.

Gold long-finned danio / *Brachydanio rerio*

The gold long-finned danio has been specially bred from the zebra danio (see page 85) and then introduced into the wild, where it has become well established in certain areas. Carefully selected adults, with attractive coloration and long fins, were bred to create a new variety of the zebra danio.

Feeding

They will eat all types of live, frozen and flake food. They swim at all levels of the aquarium and, no matter where they are, will always be one of the first to get to the food at feeding time.

ORIGIN
hybrid

TEMPERATURE RANGE
22–26°C (72–79°F)

COMMUNITY
excellent

ADULT SIZE
female 5cm (2in.)
male 5cm (2in.)

DIET
all foods

EASE OF KEEPING
10/10

pH RANGE
6.8–7.5

White-cloud mountain minnow / *Tanichthys albonubes*

ORIGIN
China

TEMPERATURE RANGE
20–26°C (68–79°F)

COMMUNITY
excellent

ADULT SIZE
female 4cm (1.5in.)
male 4cm (1.5in.)

DIET
all foods

EASE OF KEEPING
10/10

pH RANGE
6.8–7.5

Once settled and established in an aquarium, this is a really hardy fish, and is ideal for a new aquarium. It will accept a wide variety of water conditions without too many problems. It can also be kept in cooler water, but is much more active at tropical temperatures. It is a very colourful fish, with a brown-to-bronze body colour, a fluorescent-green line running the full length of the body, a bright-red tail and red fringes to the dorsal and anal fin. They look their best when swimming in a good-sized shoal. They are very easy to breed, but need dense plants so that the eggs are left alone.

Feeding

They will eat all types of live, frozen and flake food. They swim at all levels of the aquarium and, no matter where they are, will always be one of the first to get to the food at feeding time.

Siamese fighting fish / *Betta splendens*

As its name suggests, the male fish can be extremely pugnacious, but only towards males of the same species. In the Far East they have been regarded in much the same way as fighting cocks, but the sport is now banned.

In the wild, Siamese fighting fish are normally a very drab, dark green, but selective breeding can achieve a wide range of colours (reds, blues, greens and even pink). The female is the smaller of the two, with short fins and a much drabber coloration. It is advisable to place at least four or five females in an aquarium with one male so that he can divide his attentions between them.

Siamese fighting fish are bubble-nest breeders. When placed in a breeding tank, the male will build a floating bubble nest by coating air bubbles with saliva and then spitting them out. He will continue to do this for quite some time, normally until the nest measures several inches across and is up to 2.5cm (1in.) thick. When it is complete, the male courts the female until she is drawn under the nest, where he wraps his body quite tightly around hers to expel the eggs, which he then fertilises. The female sinks to the bottom of the aquarium in a semi-conscious state, while the male collects the eggs in his mouth and spits them into the protection of the bubble nest, by which time the female is ready for the next embrace. Once the female is empty of eggs, she is driven away. At this point, remove the female and place her in a small tank to recuperate from her exertions.

The male will continue to repair the nest and collect in his mouth any fry that fall out, putting them back in again. After 36 to 48 hours, the fry start hatching, and there are generally about 200 to 250 youngsters. As they hatch, they fall to the bottom of the aquarium and the male continually picks them and puts them back into the nest, so that their tails hang out. Once the fry are free-swimming, the male is unable to care for them all, so it is advisable to return him to the community aquarium. The fry have to be fed very fine food, such as infusoria and microsorium, for the first five to seven days and then progressively larger food.

Feeding

Siamese fighting fish are not a problem to feed and will eat all types of live, frozen and flake food. They swim quite slowly and are not always first in the feeding chain, so ensure that there is plenty of food for them without polluting the aquarium.

ORIGIN
Thailand

TEMPERATURE RANGE
22–26°C (72–79°F)

COMMUNITY
excellent

ADULT SIZE
female 5cm (2in.)
male 7cm (2.75in.)

DIET
all foods

EASE OF KEEPING
10/10

pH RANGE
6.8–7.5

Male fighting fish, red.

Male fighting fish, blue.

Male fighting fish, variation.

Male fighting fish, variation.

Male fighting fish, variation.

Male fighting fish, variation.

Female fighting fish.

Female fighting fish, variation.

Honey gourami / *Colisa chuna*

ORIGIN
Asia

TEMPERATURE RANGE
22–26°C (72–79°F)

COMMUNITY
excellent

ADULT SIZE
female 4cm (1.5in.)
male 4cm (1.5in.)

DIET
all foods

EASE OF KEEPING
10/10

pH RANGE
6.8–7.5

An exceptionally peaceful fish, the honey gourami is a dwarf by comparison with most gouramies. It is an ideal fish for the community aquarium as it is hardy, adaptable and causes no problems to its tank mates. It likes plenty of bushy plants in the aquarium and can be shy until well established. The male has a honey-gold body, with a black bar running from the snout down the underneath of the body and including the first half of the anal fin. The honey-gold colour covers the remainder of the anal fin and all of the dorsal fin. The tails starts from the body with this coloration, but the rear half is clear. If the fish is frightened, the black bar will vanish very quickly and can take several hours to return. The female is very plain by comparison.

Honey gourami, male.

Honey gourami, female.

Feeding

Generally, the honey gourami is not a problem to persuade to feed. However, the food does need to be quite small, so if feeding flake food, remember to break it down into much smaller flakes.

Red honey gourami, male.

Red honey gourami, female.

Dwarf gourami / *Colisa lalia*

The male dwarf gourami is an absolutely striking fish. With its many bright-red lines, made up of small spots running from the top to the bottom of its body, and fins of metallic blue, with red spotting all over them, there is every reason to have it in your aquarium. There is also a sky-blue cheek patch on both sides of its head. The ventral fins on this fish are used as feelers, sensing what it touches and they are used constantly. The female is silvery in colour and has very faint lines of blue running down her body, so they are very easy to identify. Breeding is very similar to the Siamese fighting fish (see page 91), except that the male is not quite as aggressive. Males of this species will co-exist without any problems.

ORIGIN
China

TEMPERATURE RANGE
22–26°C (72–79°F)

COMMUNITY
excellent

ADULT SIZE
female 4cm (2in.)
male 6cm (2.3in.)

DIET
all foods

EASE OF KEEPING
10/10

pH RANGE
6.8–7.5

Dwarf gourami, male.

Feeding

The dwarf gourami is not an aggressive feeder, but is very dainty in its eating habits. You must ensure that there is enough food in the aquarium for it to receive its share without polluting the tank. It will accept most foods offered to it.

Dwarf gourami, female.

Neon dwarf gourami / *Colisa lalia*

The pictures shown on pages 101-102 are of the neon-red dwarf gourami and the neon-blue dwarf gourami, colour variants of the standard dwarf gourami (see pages 99 to 100). If you care to breed them, you will find that they breed true to their own coloration. These colour variations have been bought about by demand from fish-keepers, and commercial breeders, mainly in the Far East, do their best to satisfy this demand.

ORIGIN
China

TEMPERATURE RANGE
22–26°C (72–79°F)

COMMUNITY
excellent

ADULT SIZE
female 5cm (2in.)
male 6cm (2.3in.)

DIET
all foods

EASE OF KEEPING
10/10

pH RANGE
6.8–7.5

Neon-red dwarf gourami, male.

Feeding

The neon dwarf gourami is not an aggressive feeder, but is very dainty in its eating habits. You must ensure that there is enough food in the aquarium for them. They will accept most foods offered to them.

Neon-red dwarf gourami, female.

Neon-blue dwarf gourami, male.

Neon-blue dwarf gourami, female.

Pink kissing gourami / *Helostoma temminkii*

ORIGIN
South America

TEMPERATURE RANGE
22–26°C (72–79°F)

COMMUNITY
very good

ADULT SIZE
female 30cm (12in.)
male 30cm (12in.)

DIET
mainly vegetables

EASE OF KEEPING
9/10

PH RANGE
6.8–7.5

This fish can become very large given the right conditions; in a normal aquarium they grow to about 15cm (6in.). They grow very slowly and will generally live for quite a long time. Body coloration is normally a creamy pink all over, with clear fins. It is hard to tell the difference between male and female. A spawning pair will lay literally thousands of eggs, but it is very difficult to raise the fry. They get their name from 'kissing' each other, a test of strength and willpower rather than affection. They are an ideal community fish until they grow too big for the aquarium.

Feeding

Kissing gouramis will eat some of the softer and tastier plants in your aquarium and will also rasp algae off the tank walls and harder-leafed plants. Ensure that they receive plenty of foods containing algaes, spirulina and other vegetable additives. They will feed four or five times a day.

Pearl gourami / *Trichogaster leeri*

ORIGIN
Asia

TEMPERATURE RANGE
22–26°C (72–79°F)

COMMUNITY
excellent

ADULT SIZE
female 11cm (4.3in.)
male 11cm (4.3in.)

DIET
all foods

EASE OF KEEPING
10/10

pH RANGE
6.8–7.5

Pearl gourami have also been known as lace gourami, leeri gourami and the mosaic gourami. Although they grow to a good size, they are an extremely peaceful addition to an aquarium. They are certainly very colourful fish. They have a golden body colour, with a black line running from the snout through the eye that stops three-quarters of the way along the length of the body. There is a fine spotting over the whole of the body. As the male becomes mature, his dorsal fin extends to the point where it is hanging over the tail, and the anal fin has single ray extensions. He also develops a brilliant-red overtone in the chest area. The female is also very colourful, but her fins have rounded ends.

Feeding

Pearl gouramis are quite easy to feed, taking most foods offered to them. They will quite happily take a diet of live, frozen and flake food. They also enjoy occasional supplements of mosquito larvae.

Three-spot gourami / *Trichogaster trichopterus*

ORIGIN
Asia

TEMPERATURE RANGE
22–26°C (72–79°F)

COMMUNITY
excellent

ADULT SIZE
female 11cm (4.3in.)
male 11cm (4.3in.)

DIET
all foods

EASE OF KEEPING
10/10

pH RANGE
6.8–7.5

The *Trichogaster* group of fish are very closely related to each other, but are diverse in their coloration and markings. The three-spot gourami grows to a respectable size and is a good community fish. The whole body is a faded blue, with a slightly darker shade of blue in a rippling effect on top of this. The male has elongated, pointed dorsal and anal fins, whereas the female's fins are shorter and rounded. They also have two pronounced spots, the eye being regarded as the third, hence the name. The fish are quite hardy with regard to water conditions. They like a well-planted aquarium and, because of their size, larger plants would be preferable to smaller ones. They are quite easy to breed, but must spawn in a separate aquarium.

Three-spot gourami.

Three-spot gourami, blue variation.

Feeding

Three-spot gouramis will happily take a diet of live, frozen and flake food. They also enjoy occasional supplements of mosquito larvae.

Blue or Cosby gourami / *Trichogaster trichopterus*

ORIGIN
Asia

TEMPERATURE RANGE
22–26°C (72–79°F)

COMMUNITY
excellent

ADULT SIZE
female 11cm (4.3in.)
male 11cm (4.3in.)

DIET
all foods

EASE OF KEEPING
10/10

pH RANGE
6.8–7.5

There are three main colour varieties of *Trichogaster trichopterus*. This fish, the blue or Cosby gourami, is very similar to the three-spot gourami (see pages 107), but has no spots on its body, while the shading of blue is much deeper and stronger. The darker blue appears as a marbled, rather than rippled, pattern as on the three-spot gourami, and is also very attractive. It is a good community fish, although if harassed by other tank mates it will give chase to warn them off, but does no damage. This group of fish are air-breathers that come to the surface to take in gulps of air, so they do not require such high levels of oxygen in their water as most other fish.

Feeding

Blue gouramis will happily take a diet of live, frozen and flake food. They also enjoy occasional supplements of mosquito larvae.

Gold gourami / *Trichogaster trichopterus*

This is probably the best-known member of the trichogasters, probably because of its strong colour. It has a yellow-to-gold coloration covering its entire body, extending into the fins and then gradually fading until the end of the fin is clear. The exception is the anal fin, where there is white spotting and, in a good specimen, a white edging running the full length of the fin. The males have extended and pointed fins, while the colour in the female is not as strong as in the male. They are very easy to breed and are ideal for the novice aquarist. The young grow to about 3.5cm(1.4in.) and gain full colour in about eight weeks.

Feeding

Gold gouramis will happily take a diet of live, frozen and flake food. They also enjoy occasional supplements of mosquito larvae.

ORIGIN
Asia

TEMPERATURE RANGE
22–26°C (72–79°F)

COMMUNITY
excellent

ADULT SIZE
female 11cm (4.3in.)
male 11cm (4.3in.)

DIET
all foods

EASE OF KEEPING
10/10

pH RANGE
6.8–7.5

Indian giant gourami / *Colisa fasciata*

ORIGIN
India

TEMPERATURE RANGE
22–26°C (72–79°F)

COMMUNITY
excellent

ADULT SIZE
female 9cm (3.5in.)
male 10cm (3.9in.)

DIET
all foods

EASE OF KEEPING
10/10

pH RANGE
6.8–7.5

Anabantoids originate in murky, oxygenless waters, and their strong colouring and markings are necessary to attract mates. This fish is typical, with a deep, thick body and strong markings. The male has a reddish-brown body colour, with about seven, bright-metallic-blue, vertical bars overlaid. The blue also runs the length of the anal fin where it joins the body, and there is also a bright-orange edging to this fin. The dorsal fin has a white edge to it and the tail is clear, with red edging and blue spotting all over. Ventral fins are bright orange. The dorsal fin is also elongated and pointed. The female is much drabber in its coloration and markings.

Indian giant gourami, male.

Indian giant gourami, female.

Feeding

As with nearly all gouramis, these fish are easy to feed on a diet of live, frozen and flake food. They also enjoy occasional supplements of mosquito larvae, a natural food for them in the wild.

Sparkling gourami / *Trichopsis pumila*

ORIGIN
Asia

TEMPERATURE RANGE
22–26°C (72–79°F)

COMMUNITY
excellent

ADULT SIZE
female 4cm (1.5in.)
male 4cm (1.5in.)

DIET
all foods

EASE OF KEEPING
10/10

pH RANGE
6.8–7.5

This beautiful fish is often overlooked because it is so small. Once established in an aquarium, the body colour is a very plain, silvery colour, with reddish-brown spots running the full length of the body; there are deep shades of blue and reds in the fins, with brilliant white in the ventral fins. All the body scales are reflective, and under good aquarium lighting this fish literally does sparkle.

Feeding

Sparkling gouramis will happily take a diet of live, frozen and flake food. They have small mouths, so make sure that their food is small enough for them to eat.

Thicklip gourami / *Colisa labiosa*

A very close relative of the Indian giant gourami (see pages 110 to 111), the thicklip gourami is a species in its own right. Although bodily the same shape, the thicklip gourami does not grow to quite to the same size as the Indian giant gourami. The male normally has a chocolate-brown body, with metallic-blue bars, but they are closer together than the Indian giant gourami markings. The lips are dark blue, which accentuates their size. The female is much lighter in her coloration. When in breeding condition, the male's colour changes almost to black. Both fish swim with their ventral fins pointed in front of them to sense what is in the water.

Feeding

As with nearly all gouramis, they are easy to feed on a diet of live, frozen and flake food. They also enjoy occasional supplements of mosquito larvae, a natural food for them in the wild.

ORIGIN
asia/India

TEMPERATURE RANGE
22–26°C (72–79°F)

COMMUNITY
excellent

ADULT SIZE
female 7cm (2.8in.)
male 8cm (3.1in.)

DIET
all foods

EASE OF KEEPING
10/10

pH RANGE
6.8–7.5

Thicklip gourami, female.

Thicklip gourami, male.

Angelfish / *Pterophyllum scalare*

Angelfish have acquired a poor reputation for chasing and eater smaller fish. Until they are about 7cm (2.8in.) long, they are harmless, but if they are not well fed once they are bigger, they prey upon smaller fish. For all this, they are very beautiful, with many colour and pattern variations.

Angelfish are normally sold at about 3cm(1.2in.) long. They are a very hardy fish and will tolerate a wide range of water conditions. Many species have long fins, so there must be no sharp objects in the aquarium, such as spa gravel, or fin-nipping fish, otherwise the fins will become ragged and prone to diseases like fungus and fin-rot. Different varieties of angelfish mix together happily in the same aquarium.

Angelfish will readily lay eggs, but hatching them to grow on is more difficult. Immature breeding pairs often eat their eggs within 24 hours, so remove eggs to a separate breeding tank after spawning. Treat them for fungus and supply plenty of aeration, and they should hatch. If you are fortunate enough to obtain a good pair of adults, you will find that once the eggs are laid, both parents take turns in fanning the eggs, which prevents waste matter from settling on them and turning them white with fungus. Sometimes an adult lifts an infected egg and eats it so that it does not infect the other eggs.

Black marble angelfish.

ORIGIN
South America

TEMPERATURE RANGE
22–26°C (72– 79°F)

COMMUNITY
very good when small

ADULT SIZE
female 10cm (4in.)
male 10cm (4in.)

DIET
all foods

EASE OF KEEPING
9/10

pH RANGE
6.8–7.5

Feeding

Generally, these fish are very easy to feed. Their mouths distend and open quite wide. They will accept large food that should include a diet of live, frozen and flake food. They will also accept chopped earthworm and shredded beef heart.

Gold long-fin angelfish.

Silver angelfish.

Leopard long-fin angelfish.

Zebra long-fin angelfish.

Goldhead marble angelfish.

Half-black angelfish.

Koi angelfish.

Silver diamond angelfish.

Discus fish / *Symphysodon sp.*

ORIGIN
South America

TEMPERATURE RANGE
22–26°C (72–79°F)

COMMUNITY
excellent

ADULT SIZE
female 15cm (6in.)
male 15cm (6in.)

DIET
all foods

EASE OF KEEPING
8/10

pH RANGE
6.8–7.5

Discus fish are among the most diverse of fish, being available in a wide variety of colours and patterns. They are not cheap, ranging from just a few pounds to as much as £300 to £400. They grow to a good size, but are excellent community fish and co-exist with cardinal tetras (see page 37), neon tetras (see page 36) or rummynose tetras (see page 38). Sexing is very difficult, so it is advisable to buy a number of young fish and hope that you have a pair to breed from.

Feeding

It may take a few days for them to feed once purchased, but once they are settled there should be no problems. Their mouths distend and open quite wide. They will accept large food that should include a diet of live, frozen and flake food. They will also accept chopped earthworm and shredded beef heart.

Red-turquoise discus fish.

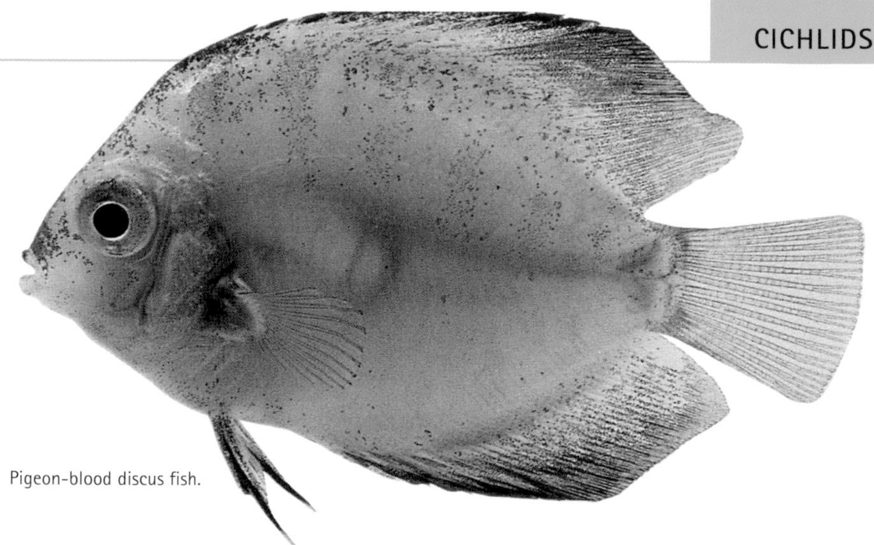

Pigeon-blood discus fish.

Blue discus fish.

Rams / butterfly cichlid / *Microgeophagus ramarezi*

ORIGIN
South America

TEMPERATURE RANGE
22–26°C (72–79°F)

COMMUNITY
excellent

ADULT SIZE
female 7cm (2.75in.)
male 7cm (2.75in.)

DIET
all foods

EASE OF KEEPING
9/10

pH RANGE
6.8–7.5

There are three different forms of this fish, all splendid additions to an aquarium. Generally known as dwarf cichlids, they have a very stocky body, a large, erect dorsal fin and many different colours on their body and fins. The base body colour is grey-blue, with black, yellow and gold on top. There is a bright-red ring around the eye and a black, vertical bar through it. The pelvic fins are mainly bright red, as are the upper and lower lobes of the tail. Sexing is quite easy, as the first few rays of the male's dorsal fin are extended. Females tends to have a red hue around the ventral area and are slightly plumper. They prefer to have a hideaway where they can swim to if they feel disturbed.

Feeding

Rams will eat all types of live, frozen and flake food. Because they swim in the lower half of the aquarium, they are not among the first to feed, so ensure that they get enough to eat without overfeeding.

Rams, standard, female.

Rams, blue, male.

Rams, blue, female.

Rams, gold, male.

Rams, gold, female.

Agassizis dwarf cichlid / *Apistogramma agassizi*

There are a number of different colour forms of this fish, some quite drab and others brilliant in their coloration. This particular variant has a golden-brown hue, with a dark bar running centrally the full length of the body. There is a black band at the base of the dorsal fin running along the edge of the body, with the remainder of the dorsal being bright orange. The anal fin is also orange and black, and the tail has a black spike coming outwards from the body, surrounded by bright orange; the edges have a line of faded black to them. The female is similar, but not quite as showy and somewhat smaller than the male. Caves or small rock formations are required by these fish, mainly for breeding and as a place of security if they feel stressed.

ORIGIN
South America

TEMPERATURE RANGE
22–26°C (72–79°F)

COMMUNITY
excellent

ADULT SIZE
female 5cm (2in.)
male 7cm (2.75in.)

DIET
all foods

EASE OF KEEPING
9/10

pH RANGE
6.8–7.5

Apistogramma agassizi, male.

Feeding

They will eat all types of live, frozen and flake food. Because they swim in the lower half of the aquarium, they are not among the first to feed, so ensure that they get enough to eat without overfeeding. Bloodworms, tubifex and other meaty foods should be included in their diet.

Apistogramma agassizi, female.

Cockatoo dwarf cichlid / *Apistogramma cacatoides*

The cockatoo dwarf cichlid has many location and colour varieties, although within the taxonomic structure they are all the same fish. The cockatoo dwarf cichlid is renowned because the first few rays of the dorsal fin are very high. All varieties of this fish have very strong colours, such as bright red, orange and jet black, in their markings. The female is normally very plain by comparison with the male and is also considerably smaller. However, when spawning, the female will resist the advances of the male until she is ready for him. They need a cave or flowerpot to spawn in and will both care for the eggs until they have hatched. Breeding pairs normally continue this care for the first two to three weeks.

ORIGIN
South America

TEMPERATURE RANGE
22–26°C (72–79°F)

COMMUNITY
excellent

ADULT SIZE
female 5cm (2in.)
male 7cm (2.75in.)

DIET
all foods

EASE OF KEEPING
9/10

pH RANGE
6.8–7.5

Apistogramma cacatoides, male.

Feeding

They will eat all types of live, frozen and flake food. Because they swim in the lower half of the aquarium, they are not among the first to feed, so ensure that they get enough to eat without overfeeding. Bloodworms, tubifex and other meaty foods should be included in their diet.

Apistogramma cacatoides, female.

Nijsseni's dwarf cichlid / *Apistogramma nijsseni*

Ideal for the novice aquarist, Nijsseni's dwarf cichlid is exceptionally easy to keep. They do not become as large as other dwarf cichlids, but make up for that with their multiple-pastel coloration. The male is normally a slate-grey blue, with slightly darker patches and a dark spot in the caudal peduncle. The ventral and anal fins are lemon, the tail has a light-blue-grey inner edging, with a light-orange outer edge, and the dorsal has two lines of deep red and very light blue running along the top. The female is slightly smaller and much weaker in coloration. They require some cover, as they can be a little shy, and also for breeding. Spawning is both difficult and erratic, but they are beautiful fish that are easy to keep, as well as maintain.

ORIGIN
South America

TEMPERATURE RANGE
22–26°C (72–79°F)

COMMUNITY
excellent

ADULT SIZE
female 5cm (2.75in.)
male 6cm (2.4in.)

DIET
all foods

EASE OF KEEPING
9/10

pH RANGE
6.8–7.5

Apistogramma nijsseni, male.

Feeding

They will eat all types of live, frozen and flake food. Because they swim in the lower half of the aquarium, they are not among the first to feed, so ensure that they get enough to eat without overfeeding. Bloodworms, tubifex and other meaty foods should be included in their diet.

Apistogramma nijsseni, female.

Vietjita dwarf cichlid / *Apistogramma vietjita II*

These are a particularly thickset and deep-bodied fish, ideal for the novice aquarist because they are so hardy and acclimatise easily to a new environment. Colour is very strong in good specimens. The reds and oranges in the fins are very bright, accentuating their size. The females are smaller and much drabber than the males. The true beauty of this species is not manifest until the fish is mature, and as only immature specimens are sold, it is not particularly popular. Breeding is easy.

ORIGIN
South America

TEMPERATURE RANGE
22–26°C (72–79°F)

COMMUNITY
excellent

ADULT SIZE
female 5cm (2.75in.)
male 6cm (2.4in.)

DIET
all foods

EASE OF KEEPING
9/10

pH RANGE
6.8–7.5

Apistogramma vietjita, male.

Feeding

They will eat all types of live, frozen and flake food. Because they swim in the lower half of the aquarium, they are not among the first to feed, so ensure that they get enough to eat without overfeeding. Bloodworms, tubifex and other meaty foods should be included in their diet.

Apistogramma vietjita, female.

Keyhole cichlids / *Aequidens maronii*

With a brown body colour, this is not one of the most colourful fish, although it is certainly one of the most peaceful cichlids. Cichlids have a reputation for damaging aquarium plants, but the keyhole cichlid is an exception. They grow to become very stocky and thick-bodied. Sexing is very difficult, except when the female is full of roe and ready to spawn. These fish do not breed on a regular basis. Spawning will occur two or three times and will then stop, maybe for some months. When they have spawned and the fry have hatched out, you will find that they are quite easy to rear, as they grow to 3cm (1in.) in eight to ten weeks.

Feeding

These fish will eat just about all types of live, frozen and flake food, as well as finely grated beef heart. They seem to feed very delicately and are slow in the way that they feed.

ORIGIN
South America

TEMPERATURE RANGE
22–26°C (72–79°F)

COMMUNITY
very good

ADULT SIZE
female 9cm (3.5in.)
male 10cm (4 in.)

DIET
all foods

EASE OF KEEPING
10/10

pH RANGE
6.8–7.5

Thomas's dwarf Cichlid / *Pelmatochromis thomasi*

ORIGIN
Africa

TEMPERATURE RANGE
22–26°C (72–79°F)

COMMUNITY
excellent

ADULT SIZE
female 8cm (3.2in.)
male 8cm (3.2in.)

DIET
all foods

EASE OF KEEPING
10/10

pH RANGE
6.8–7.5

Thomas's dwarf cichlid is an old favourite because of its peacefulness and beauty. It is also one of a small group of cichlids that can be placed in a community tank. They have large dorsal fins that, when extended, show yellow and red lines to the edge, with faint white patterning in the remainder. There is also a red edging to the tail fin. The body is very stocky and deep, with reflective scaling where the colour changes as the fish turns in the light. These fish are typical egg-layers and are quite easy to breed. They will clean a site before laying their eggs, with both parents taking care until the eggs have hatched and the youngsters are free-swimming.

Feeding

They will eat all types of live, frozen and flake food. Because they swim in the lower half of the aquarium, they are not among the first to feed, so ensure that they get enough to eat without overfeeding.

Lemon cichlid / *Neolamprologus leleupi*

Originally from the African lakes, this fish's normal habitat is large, rocky outcrops. In an aquarium, they prefer plenty of rocks to hide, and possibly breed, in. You can also use terracotta flowerpots. The coloration of this fish is unusual, in that it is totally bright yellow, including its fins, and so really stands out in an aquarium. The dorsal and anal fin stretch nearly the full length of the body. The body is long and tubular in shape, which suggests that it is a fast swimmer and also allows it to escape through small gaps in the rockwork if it feels threatened.

Feeding

These fish will eat just about all types of live, frozen and flake food. They will feed quite voraciously and take three or four feeds a day without any problems. They require a very varied diet, which includes free-swimming, live foods, such as daphnia, cyclops and glassworm, if available.

ORIGIN
Africa

TEMPERATURE RANGE
22–26°C (72–79°F)

COMMUNITY
very good

ADULT SIZE
female 10cm (4in.)
male 10cm (4in.)

DIET
all foods

EASE OF KEEPING
10/10

pH RANGE
6.8–7.5

Network corydoras / *Corydoras agassizi*

ORIGIN
South America

TEMPERATURE RANGE
22–26°C (72–79°F)

COMMUNITY
excellent

ADULT SIZE
female 7cm (2.75in.)
male 7cm (2.75in.)

DIET
all foods

EASE OF KEEPING
10/10

pH RANGE
6.8–7.5

With a silver body and a black pattern, this is a fish that stands out well in an aquarium. The base colour of the body is a very bright, silvery white, with four black lines broken up into unevenly edged, small blotches. There is a large, jet-black blotch in the base of the dorsal fin that runs into the body. The fins are clear, and the tail has a fine pattern of black vertical lines in it. *Corydoras agassizi* is a very hardy fish that can be introduced into a relatively new aquarium after about four weeks. A small shoal of these fish is preferable, as they like to swim across the substrate, grazing for food together. Breeding is quite easily achieved, given the right conditions.

Feeding

As with other corydoras, they will eat just about all live, frozen and flake food. They swim in the lower half of the aquarium and are not one of the first to feed, so ensure that there is enough for them to eat without overfeeding.

Albino corydoras / *Corydoras aeneus*

This is one of the three most common corydoras available, partly because it is so easy to breed. The albino corydoras is one of a number of fish that are completely supplied by commercial breeders (none are captured from the wild). They are an extremely hardy fish and can be placed in a two-week-old aquarium. This fish is an albino form of the bronze corydoras (see page 139). Not only is the eye reddish pink, but the body is pink all over. Sexing is quite easy: when looking from above, the female appears much more barrel-shaped than the male. Use only rounded gravel, such as Dorset pea, which will not ruin their long barbels. As with all corydoras, a sand substrate would be better.

Feeding

As with other corydoras, they will eat just about all live, frozen and flake food. They swim in the lower half of the aquarium and are not one of the first to feed, so ensure that there is enough for them to eat without overfeeding.

ORIGIN
South America

TEMPERATURE RANGE
22–26°C (72–79°F)

COMMUNITY
excellent

ADULT SIZE
female 7cm (2.75in.)
male 7cm (2.75in.)

DIET
all foods

EASE OF KEEPING
10/10

pH RANGE
6.8–7.5

Arched corydoras / *Corydoras arcuatus super*

ORIGIN
South America

TEMPERATURE RANGE
22–26°C (72–79°F)

COMMUNITY
excellent

ADULT SIZE
female 10cm (4in.)
male 10cm (4in.)

DIET
all foods

EASE OF KEEPING
10/10

pH RANGE
6.8–7.5

The arched corydoras is a very attractive and hardy fish. It has a stocky body, with a creamy-white base body colour and a thick, black bar running through the eye, following the crown of the back, right through to the lower lobe of the tail. This black bar will practically disappear when the fish is stressed or frightened. Once settled, it will reappear quite quickly. The snout is extremely rounded, with prominent barbels. Sexing is reasonably easy, but breeding can be a challenge. Once conditions are right, the fry are quite easy to raise.

Feeding

As with other corydoras, they will eat just about all live, frozen and flake food. They swim in the lower half of the aquarium and are not one of the first to feed, so ensure that there is enough for them to eat without overfeeding.

Bronze corydoras / *Corydoras aeneus*

This is one of the three most common corydoras available, partly because it is so easy to breed. Almost all bronze corydoras are supplied by commercial breeders in the Far East and are not collected from wild habitats. They are extremely hardy, although they can be shy and hide in plants. As their name suggests, they are bronze brown, normally with two very dark, large patches in the body. This is an ideal fish to add to a two-week-old aquarium. Use only rounded gravel, such as Dorset pea, because this will not ruin their long barbels. As with all corydoras, a sand substrate would be better.

Feeding

They will eat just about all live, frozen and flake food. They swim in the lower half of the aquarium and are not one of the first to feed, so ensure that there is enough for them to eat without overfeeding.

ORIGIN
South America

TEMPERATURE RANGE
22–26°C (72–79°F)

COMMUNITY
excellent

ADULT SIZE
female 8cm (3.2in.)
male 8cm (3.2in.)

DIET
all foods

EASE OF KEEPING
10/10

pH RANGE
6.8–7.5

Elegant corydoras / *Corydoras elegans*

ORIGIN
South America

TEMPERATURE RANGE
22–26°C (72–79°F)

COMMUNITY
excellent

ADULT SIZE
female 7cm (2.75in.)
male 7cm (2.75in.)

DIET
all foods

EASE OF KEEPING
10/10

pH RANGE
6.8–7.5

These are fish that have been known and kept by aquarists for many years. They do not grow as large as many other corydoras and so suit slightly smaller aquariums. They are widely available and are reasonably cheap. They are very easy to keep and maintain. Sexing is easy: the male is quite slim, and from above the female is more rotund. Breeding and raising is simple, as there is no parental care from the adults. It is best to spawn them in a separate aquarium, where the fry can be raised.

Feeding

They will eat just about all live, frozen and flake food. They swim in the lower half of the aquarium and are not one of the first to feed, so ensure that there is enough for them to eat without overfeeding.

Pygmy corydoras / *Corydoras habrosus*

Corydoras habrosus is one of a small group of true pygmy corydoras. When first imported, they are no more than about 1.5 to 2cm (0.5 to 0.75in.) in length. In the wild, they shoal in great numbers, so the more you can place in your aquarium, the happier they will be. They tend to swim and hover together just off the bottom of the aquarium rather than grazing the substrate looking for food like most other corydoras. They are a very attractive little fish, with a barrel-like body and pretty markings. They are very easy to sex, as the female is much thicker in the body than the male, but they are quite difficult to breed.

Feeding

They will eat just about all live, frozen and flake food. They swim in the lower half of the aquarium and are not one of the first to feed, so ensure that there is enough for them to eat without overfeeding.

ORIGIN
South America

TEMPERATURE RANGE
22–26°C (72–79°F)

COMMUNITY
excellent

ADULT SIZE
female 4cm (1.5in.)
male 4cm (1.5in.)

DIET
all foods

EASE OF KEEPING
10/10

pH RANGE
6.8–7.5

Leopard corydoras / *Corydoras jullii*

ORIGIN
South America

TEMPERATURE RANGE
22–26°C (72–79°F)

COMMUNITY
excellent

ADULT SIZE
female 7cm (2.75in.)
male 7cm (2.75in.)

DIET
all foods

EASE OF KEEPING
10/10

pH RANGE
6.8–7.5

The body of the leopard corydoras is covered in tiny black spots, with a line of slightly larger spots running down the lateral line of the fish over a base colour of silvery white, hence the name of the fish. The dorsal has a large black blotch in the upper part, and the tail has a fine pattern all over it. It is a very pretty fish. The snout is extremely rounded and blunt. *Corydoras jullii* is one of the most common corydoras available. When breeding, this fish will quite happily cross with *Corydoras trilineatus*, which will achieve a slight variation in patterning. They are very easy to keep and maintain, and like well-oxygenated and well-filtered aquarium water.

Feeding

They will eat just about all live, frozen and flake food. They swim in the lower half of the aquarium and are not one of the first to feed, so ensure that there is enough for them to eat without overfeeding.

False-bandit corydoras / *Corydoras melini*

This is a recent addition to aquaria, but is now well established. Not as freely available as some, it is slightly more expensive than many other corydoras. It is a very good-looking, solid fish with a creamy base body colour and a black bar running from within the dorsal fin, down the crown of the back into the lower lobe of the tail. There is also a wide black bar running vertically through the eye, which is known as a mask, hence its name. It is quite a hardy fish that will accept small changes in pH levels, but does need high oxygenation levels and a well-maintained filtration system. It will not tolerate any nitrite problems.

Feeding

They will eat just about all live, frozen and flake food. They swim in the lower half of the aquarium and are not one of the first to feed, so ensure that there is enough for them to eat without overfeeding.

ORIGIN
South America

TEMPERATURE RANGE
22–26°C (72–79°F)

COMMUNITY
excellent

ADULT SIZE
female 6cm (2.4in.)
male 6cm (2.4in.)

DIET
all foods

EASE OF KEEPING
10/10

pH RANGE
6.8–7.5

Peppered corydoras / *Corydoras paleatus*

ORIGIN
South America

TEMPERATURE RANGE
22–26°C (72–79°F)

COMMUNITY
excellent

ADULT SIZE
female 7cm (3in.)
male 7cm (3in.)

DIET
all foods

EASE OF KEEPING
10/10

pH RANGE
6.8–7.5

This fish is one of the three most popular corydoras available. Commercial breeders supply most of them because they are so easy to breed. They are extremely peaceful, hardy and long-lived, and have a wonderful habit of sitting on the substrate looking through the aquarium's front glass and 'winking' at whoever is watching them. The body colour is olive green, with mottled dark-green, almost black, markings. No two fish have exactly the same patterns. The tail has faint spots and the other fins are usually clear of any colour. The male has a much larger, pointed dorsal fin than the female, although body shape is a clearer aid to identification.

Feeding

They will eat just about all live, frozen and flake food. Live food, such as tubifex and bloodworms, is ideal for this type of bottom-scavenging fish, especially if they are on a sand substrate.

Spotted or dotted corydoras / *Corydoras punctatus*

You can buy so many pattern variations of this fish that it is very difficult to give a concise and accurate description. Generally, the dorsal fin has a large, black spot, the body has spots all over it (ranging from tiny dots to large spots), the tail has fine, spotted markings, the other fins are clear and the snout is rounded. Basically, they all have a black-and-white pattern and the variations are very pretty. As with all other corydoras, they are community fish, seemingly always grubbing around the bottom of the tank, looking for food. They are one of the easier corydoras to breed. They lay adhesive eggs, normally on the glass wall of the aquarium, which stick there until they hatch and then drop to the bottom of the aquarium. There is no parental care from the adults.

Feeding

They will eat just about all live, frozen and flake food. Live food, such as tubifex and bloodworms, is ideal for this type of bottom-scavenging fish, especially if they are on a sand substrate.

ORIGIN
South America

TEMPERATURE RANGE
22–26°C (72–79°F)

COMMUNITY
excellent

ADULT SIZE
female 7cm (3in.)
male 7cm (3in.)

DIET
all foods

EASE OF KEEPING
10/10

pH RANGE
6.8–7.5

Flag-tail corydoras / *Corydoras robbinae*

ORIGIN
South America

TEMPERATURE RANGE
22–26°C (72–79°F)

COMMUNITY
excellent

ADULT SIZE
female 8cm (3.2in.)
male 8cm (3.2in.)

DIET
all foods

EASE OF KEEPING
10/10

pH RANGE
6.8–7.5

This is an active, eye-catching fish, constantly on the move and with striking black-and-white markings on the tail. It is not a shy fish that retires amongst the plants, but is nearly always at the front of the aquarium and on show. There is very little information about breeding. The male and female are identical in markings, but the female, as with most other corydoras, is plumper in the body.

Feeding

As with other corydoras, they will eat just about all live, frozen and flake food. They swim in the lower half of the aquarium and are not one of the first to feed, so ensure that there is enough for them to eat without overfeeding.

Schwartz's corydoras / *Corydoras schwartzi*

There are so many pattern variations of this fish that it is very difficult to give an accurate description. Generally, there is a large, black bar running up the front of the dorsal fin. The body has lines of broken spots along it, ranging from tiny dots to large spots, the tail has fine, spotted markings, the other fins are clear and the snout is rounded. The barbels are white and quite pronounced. Basically, they all have a black-and-white pattern. All of the variations are very pretty. As with all other corydoras, they are good community fish, often grubbing around the bottom of the tank, looking for food. They are one of the easier corydoras to breed. They lay adhesive eggs, normally on the glass wall of the aquarium, where they stick until they hatch, when they drop to the bottom of the aquarium. There is no parental care from the adults.

Feeding

As with other corydoras, they will eat just about all live, frozen and flake food. They swim in the lower half of the aquarium and are not one of the first to feed, so ensure that there is enough for them to eat without overfeeding.

ORIGIN
South America

TEMPERATURE RANGE
22–26°C (72–79°F)

COMMUNITY
excellent

ADULT SIZE
female 7cm (3in.)
male 7cm (3in.)

DIET
all foods

EASE OF KEEPING
10/10

pH RANGE
6.8–7.5

Sterbai's corydoras / *Corydoras sterbai*

ORIGIN
South America

TEMPERATURE RANGE
22–26°C (72–79°F)

COMMUNITY
excellent

ADULT SIZE
female 7cm (3.2in.)
male 7cm (3.2in.)

DIET
all foods

EASE OF KEEPING
10/10

pH RANGE
6.8–7.5

This fish is another relative newcomer to the hobby, and a very beautiful one at that. *Corydoras sterbai* is a very stocky fish, being quite wide in the body, as well as deep, with striking patterning. It has very strong lines down the length of the body, made up of a series of large dots, with patterns on the fins, too. They have long barbels so that they can dig around in the substrate for food. The snout is very blunt and rounded. This fish is very similar to *Corydoras haroldschultzi* and is easily confused with it. Little is known about breeding this fish, but it is presumed to be the same as other corydoras.

Feeding

As with other corydoras, they will eat just about all live, frozen and flake food. They swim in the lower half of the aquarium and are not one of the first to feed, so ensure that there is enough for them to eat without overfeeding.

Plecostomus / *Hypostomus plecostomus*

The plecostomus is like something out of the age of dinosaurs, with large, hard scales and consequently few enemies. They will attach themselves to the glass walls of an aquarium, or to ornaments or bogwood, grazing for any algae that may have grown there. Any plants will need to be strong and hardy. The plecostomus grows quite large, but is not a threat to other fish; it does not attach itself to the sides of other fish unless that fish is severely injured or dying. The main problem, however, is that it can outgrow the aquarium.

Feeding

Although they will feed on any algae in an aquarium, they do have specific feeding requirements. They will feed on just about any food, but prefer scalded lettuce leaves, boiled garden peas and spinach. The spinach has to be sugar-free and is best from a tin.

ORIGIN
South America

TEMPERATURE RANGE
22–26°C (72–79°F)

COMMUNITY
excellent

ADULT SIZE
female 30cm (12in.)
male 30cm (12in.)

DIET
mainly vegetables

EASE OF KEEPING
10/10

pH RANGE
6.8–7.5

Sail-fin pleco / red gibbiceps / *Pterygoplichthys gibbiceps*

ORIGIN
South America

TEMPERATURE RANGE
22–26°C (72–79°F)

COMMUNITY
excellent

ADULT SIZE
female 30cm (12in.)
male 30cm (12in.)

DIET
mainly vegetables

EASE OF KEEPING
10/10

pH RANGE
6.8–7.5

The red gibbiceps resembles the forbidding plecostomus (see page 144), with large, hard scales on the body. It grazes aquarium walls and ornaments for algae, and although it grows very large, is a gentle giant.

Feeding

Although they will feed on any algae in an aquarium, they do have specific feeding requirements. They will feed on just about any food, but prefer scalded lettuce leaves, boiled garden peas and spinach. The spinach has to be sugar-free and is best from a tin.

L204 flash peckoltia / *L204 Flash peckoltia*

ORIGIN
South America

TEMPERATURE RANGE
22–26°C (72–79°F)

COMMUNITY
excellent

ADULT SIZE
female 15cm (6in.)
male 15cm (6in.)

DIET
mainly cockles and mussels

EASE OF KEEPING
9/10

pH RANGE
6.8–7.5

This fish has only recently been identified, hence the dentifying number, L204. When first imported, they are only about 5 to 6cm (1.9 to 2.3in.) long, but their pattern and beauty is the same at that size as when fully adult. They have a distinctive, dark-brown body and bright-white, thin striping all over the body and fins. They are quite expensive, but are well worth it for their beauty. These fish require real bogwood to hide under and graze on, rasping off very fine pieces to aid their digestion.

Feeding

Chopped frozen cockle and mussel are the main foods of this fish. They will take very small quantities of food placed in the aquarium for the other fish, but will not exist on it.

L18 gold-nugget pleco / *L18 Baryancistrus* sp.

ORIGIN
South America

TEMPERATURE RANGE
22–26°C (72–79°F)

COMMUNITY
excellent

ADULT SIZE
female 15cm (6in.)
male 15cm (6in.)

DIET
mainly cockles and mussels

EASE OF KEEPING
9/10

pH RANGE
6.8–7.5

This beautiful fish has not been completely classified as yet. With a very dark base colour to the body and fins, it has bright-yellow spots randomly placed all over it. There is also a band of bright yellow across the top of the dorsal fin and down the end of the tail. This is similar to the markings on the L81, which has fine dots of yellow. These fish require real bogwood to hide under and graze on, rasping off very fine pieces to aid their digestion.

Feeding

Chopped frozen cockle and mussel are the main foods of this fish. They will take very small quantities of food placed in the aquarium for the other fish, but will not exist on it.

L177 queen gold nugget / *L177 Baryancistrus* sp.

This is another beautiful addition to your aquarium, which, as with the other 'L'-numbered fish shown here, has not been classified as yet. This fish is very close in marking to L18 (see page 152) and L81, the main difference being that, when mature, they have much wider bands of yellow in the fins. Very little known about the breeding habits of these fish. They require real bogwood to hide under and graze on, rasping off very fine pieces to aid their digestion.

Feeding

Chopped frozen cockle and mussel are the main foods of this fish. They will take very small quantities of food placed in the aquarium for the other fish, but will not exist on it.

ORIGIN
South America

TEMPERATURE RANGE
22–26°C (72–79°F)

COMMUNITY
excellent

ADULT SIZE
female 15cm (6in.)
male 15cm (6in.)

DIET
mainly cockles and mussels

EASE OF KEEPING
9/10

pH RANGE
6.8–7.5

L262 snowy hypancistrus / *L262 Hypancistrus* sp.

ORIGIN
South America

TEMPERATURE RANGE
22–26°C (72–79°F)

COMMUNITY
excellent

ADULT SIZE
female 9cm (3.5in.)
male 9cm (3.5in.)

DIET
mainly cockles and mussels

EASE OF KEEPING
9/10

pH RANGE
6.8–7.5

Although not yet officially identified and named, this fish has already been given the common name of snowy hypancistrus. It has become a highly sought-after fish because of its very bright and distinct markings. When first imported, it is only about 3 to 4cm (1.2 to 1.6in.) long, but its pattern and beauty is the same at that size as when fully adult. Jet-black body colour and bright-white, tiny dots all over the body and fins make it very distinctive. These fish require real bogwood to hide under and graze on, rasping off very fine pieces to aid their digestion.

Feeding

Chopped frozen cockle and mussel are the main foods of this fish. They will take very small quantities of food placed in the aquarium for the other fish, but will not exist on it.

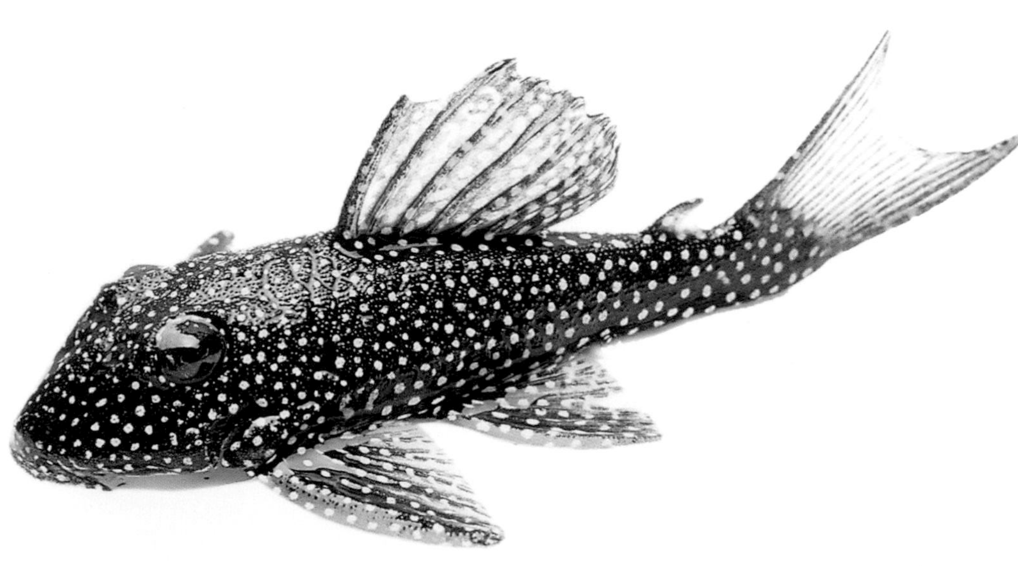

L260 queen arabesque / *L260*

Queen arabesque is now widely available and is extremely hardy and very easy to maintain once it has settled into an aquarium. Information on the sexing and breeding habits of these fish remains limited, but they have a habit of sitting on rocks or wood and just looking at you. These fish require real bogwood to hide under and graze on, rasping off very fine pieces to aid their digestion.

Feeding

Chopped frozen cockle and mussel are the main foods of this fish. They will take very small quantities of food placed in the aquarium for the other fish, but will not exist on it.

ORIGIN
South America

TEMPERATURE RANGE
22–26°C (72–79°F)

COMMUNITY
excellent

ADULT SIZE
female 9cm (3.5in.)
male 9cm (3.5in.)

DIET
mainly cockles and mussels

EASE OF KEEPING
9/10

pH RANGE
6.8–7.5

Goldspot ancistrus albino / *Ancistrus* sp. *gold albino*

ORIGIN
South America

TEMPERATURE RANGE
22–26°C (72–79°F)

COMMUNITY
excellent

ADULT SIZE
female 12cm (4.75in.)
male 12cm (4.75in.)

DIET
mainly vegetables

EASE OF KEEPING
10/10

pH RANGE
6.8–7.5

Albino fish are rare in the wild, but this fish has been commercially bred for the aquarist trade. The males have bristles running around the chin and up the centre of the head, whereas the females only have them around the edge of the chin, so sexing these fish is very easy. They have sucker mouths and will graze all over plants and the sides of the aquarium, eating the algae growth. They have sharp gill spines, so handle them very carefully.

Feeding

They will feed on just about any food that you place in the aquarium, but scalded lettuce leaves, boiled garden peas and cut spinach are favourites. The spinach has to be sugar-free and is best from a tin. Crushed cockle and chopped mussel will also be accepted.

Bristlenose ancistrus / *Ancistrus dolichopterus*

There are a number of varieties of *Ancistrus* species, but only one or two are generally available. They all have a sucker mouth underneath the head, and the males all have fleshy bristles around the chin and up the centre of the head. The females only have them around the edge of the chin, so sexing is easy. They use their mouths to secure themselves to plants and the sides of aquariums, eating the algae growth. They also have sharp gill spines, which they use defensively, so handle them carefully. When breeding, note that the eggs are yellow, instead of the usual white or opaque colour.

Feeding

These fish will feed on any algae in an aquarium, but will also eat just about any food that you place in the aquarium. Scalded lettuce leaves, boiled garden peas and cut spinach are favourites of these fish. The spinach has to be sugar-free and is best from a tin.

ORIGIN
South America

TEMPERATURE RANGE
22–26°C (72–79°F)

COMMUNITY
excellent

ADULT SIZE
female 12cm (3.75in.)
male 12cm (3.75in.)

DIET
mainly vegetables

EASE OF KEEPING
10/10

pH RANGE
6.8–7.5

Top view

Bullnose / rubbernose / *Chaetostoma thomasi*

ORIGIN
South America

TEMPERATURE RANGE
22–26°C (72–79°F)

COMMUNITY
excellent

ADULT SIZE
female 15cm (6in.)
male 15cm (6in.)

DIET
all foods

EASE OF KEEPING
9/10

pH RANGE
6.8–7.5

There are only two types of *Chaetostoma* available to the hobby: this one and the spotted bullnose, the only difference being that the bullnose does not have spots. Typical of the plecostomus-type fish, the scales are tough and rough to the touch. Their eyes are small for a fish that grows to about 15cm (6in.), but their vision is excellent. They like to graze on bogwood and the aquarium walls, and the mouth is a very efficient sucker. They also have the ability to lock their fins when feeling threatened. The sexual differences are not known.

Feeding

As with the plecostomus, they will feed on any algaes in the aquarium, as well as scalded lettuce leaves, boiled garden peas and cut spinach. The spinach has to be sugar-free and is best from a tin.

Glass or ghost catfish / *Kryptopterus bicirrhis*

Once settled and established, this fish is hardy and easy to maintain. They are shy fish and generally lurk under large-leafed plants rather than swimming freely all over the aquarium. Keeping them in a small shoal of five or six helps them to feel more secure. As the name suggests, you can see right through the body. They have two hair-like barbels that are half the length of the body, which they use as sensors. They do not have the usual-shaped dorsal fin, as theirs is made up of a single ray. Sexing and breeding of this fish are still unknown. White spot can be a minor problem, but is obvious to see and easy to treat.

Feeding

The glass catfish will accept all types of food, but prefers a wide variety of dried, frozen and live foods, with a supplement of black mosquito larvae.

ORIGIN
Asia

TEMPERATURE RANGE
22–26°C (72–79°F)

COMMUNITY
very good

ADULT SIZE
female 9cm (3.75in.)
male 9cm (3.75in.)

DIET
all foods

EASE OF KEEPING
9/10

pH RANGE
6.8–7.5

Otocinclus / *Otocinclus arnoldi*

ORIGIN
South America

TEMPERATURE RANGE
22–26°C (72–79°F)

COMMUNITY
excellent

ADULT SIZE
female 6cm (2.4in.)
male 6cm (2.4in.)

DIET
mainly vegetables

EASE OF KEEPING
10/10

pH RANGE
6.8–7.5

If you have problems with algae growth, then these fish are an absolute must: they continually feed, cleaning any algae in the tank. Because they are so small, their food source has to be constantly available, otherwise they can starve very quickly. In some aquaria, it would be advisable to put tablets that encourage algae growth into the tank. They will harm no other fish and are incompatible only with larger fish that are aggressive. When first imported or moved to a new aquarium, otocinclus can be prone to white spot, but this can easily be treated and cured. The females are only sexable when they are full of eggs and ready to breed.

Feeding

Although algaes are their main diet, they will readily accept scalded lettuce leaves, boiled peas or cut spinach. Foods fed to the other fish will be taken in small amounts, but a vegetable diet is preferable.

Red whiptail / *Loricaria* sp.

This whiptail has only been easily available for the past three or four years, but is now being commercially bred in the Czech Republic. They are a long, thin-bodied fish, being widest at the head. The tail has long filament extensions, which can be half the length of the body. This species has a red-ochre coloration all over the body, with small, darker-coloured, marbled patches on it. They are exceptionally peaceful fish, gliding, rather than swimming, around the aquarium. They like to lay and graze on large-leafed plants, such as Amazon swords. From above, the female is wider just behind the pectoral fins than the male. Breeding is possible and, when successful, the male will take care of the eggs until they have hatched.

Feeding

Algaes form their main diet, but they will readily accept scalded lettuce leaves, boiled peas or cut spinach. They will also accept other foods, such as chopped shrimp and cooked white fish.

ORIGIN
South America

TEMPERATURE RANGE
22–26°C (72–79°F)

COMMUNITY
excellent

ADULT SIZE
female 15cm (6in.)
male 15cm (6in.)

DIET
mainly vegetables

EASE OF KEEPING
9/10

pH RANGE
6.8–7.5

Tiffany whiptail / *Loricaria teffiana*

ORIGIN
South America

TEMPERATURE RANGE
22–26°C (72–79°F)

COMMUNITY
excellent

ADULT SIZE
female 18cm (7in.)
male 18cm (7in.)

DIET
mainly vegetables

EASE OF KEEPING
10/10

pH RANGE
6.8–7.5

Tiffany whiptails are now being commercially bred in the Czech Republic. They are a long, thin-bodied fish, and are widest at the head. They have broad, jet-black bands running across the body, and a much lighter band between them. The tail has very long filament extensions, which can be half the length of the body. They are exceptionally peaceful fish, gliding, rather than swimming, around the aquarium. They like to lay and graze on large-leafed plants, such as Amazon swords. From above, the female is wider just behind the pectoral fins than the male.

Feeding

Although algaes are their main diet, they will readily accept scalded lettuce leaves, boiled peas or cut spinach. Foods fed to the other fish will be taken in small amounts, but a vegetable diet is preferable.

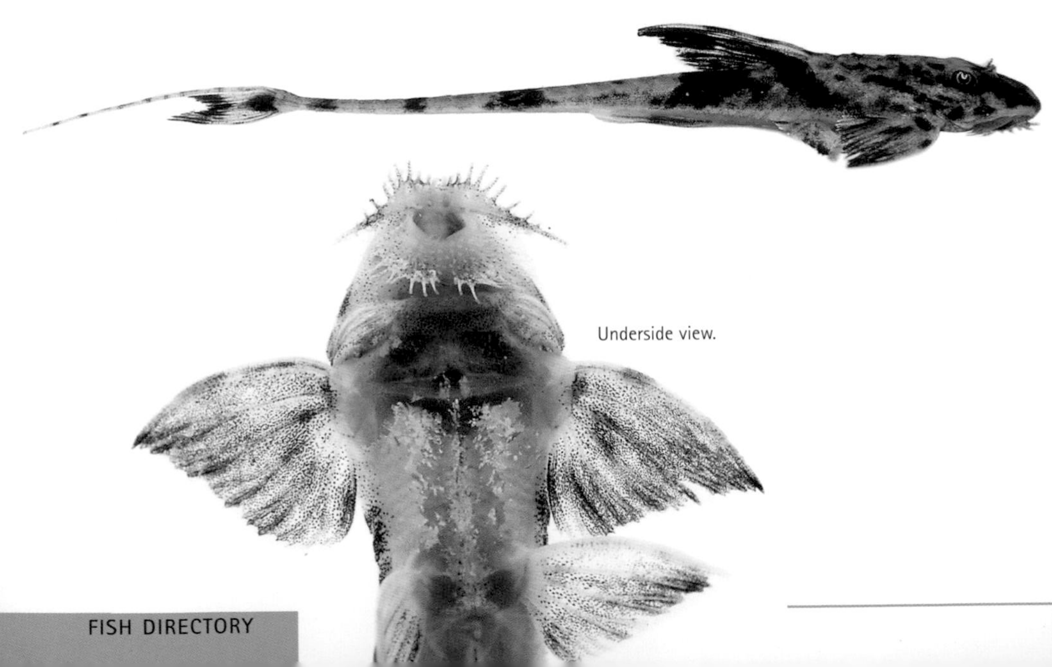

Underside view.

Twig catfish / *Farlowella acus*

Looking at this fish, it is easy to see how it derives its name: in the wild, its shape and colour act as camouflage to protect it from other predators. In the aquarium, it is extremely docile and an excellent community fish. The aquarium needs wide-leafed plants so that they can sit on them and clean the leaves of algae growth. They take a few days to settle into any new environment and will tend to hide away, but once they have acclimatised they are active and not afraid to be centrepiece of your show tank. Sexing is difficult until they are of a good, adult size, when the female will be slightly wider than the male just behind the head.

Feeding

Although algaes form their main diet, they will readily accept scalded lettuce leaves, boiled peas or cut spinach. Foods fed to the other fish will be taken in small amounts, but a vegetable diet is preferable.

ORIGIN
South America

TEMPERATURE RANGE
22–26°C (72–79°F)

COMMUNITY
excellent

ADULT SIZE
female 20cm (9in.)
male 20cm (9in.)

DIET
mainly vegetables

EASE OF KEEPING
8/10

pH RANGE
6.8–7.5

Upside-down catfish / *Synodontis nigriventris*

ORIGIN
Africa

TEMPERATURE RANGE
22–26°C (72–79°F)

COMMUNITY
very good

ADULT SIZE
female 9cm (3.5in.)
male 9cm (3.5in.)

DIET
all foods

EASE OF KEEPING
10/10

pH RANGE
6.8–7.5

As their name suggests, these fish really do swim upside down most of the time. They swim at all levels of the aquarium in this way, but will then swim in the conventional way to graze over the substrate for food. Their spines are sharp and can pierce the skin easily, so be careful when netting them. They will make croaking noises when being lifted from the water. Active fish, they are sold when they are about 3 to 4cm (1.2 to 1.6in.), but quickly grow to a more substantial size

Feeding

This fish will accept any type of food. They feed upside down from the surface, taking floating foods, or swim to the bottom of the aquarium and eat food from the substrate.

Clown loach / *Botia macracantha*

Despite their large size, clown loaches can be added to an aquarium with very small shoaling fish, such as cardinal tetras (see page 37). They are very active fish and prefer to be in a small shoal of three or four. With its strong orange-and-black coloration, this fish is certainly striking. They are prone to white spot and, as they are scaleless fish, the best way to treat this is to increase the heat in the tank. They have been successfully bred, but only in extremely large aquariums or pools.

Feeding

They require green foods in their diet, such as scalded lettuce or boiled garden peas. They are also partial to chopped shrimp, cod roe, tubifex and bloodworms. When they are about 3 to 4cm (1.2 to1.6in.) long, they need three to four small feeds a day.

ORIGIN
Asia

TEMPERATURE RANGE
22–26°C (72–79°F)

COMMUNITY
excellent

ADULT SIZE
female 20cm (8in.)
male 20cm (8in.)

DIET
mainly vegetables

EASE OF KEEPING
9/10

pH RANGE
6.8–7.5

Indian marbled loach / *Neomochielus botia*

ORIGIN
India

TEMPERATURE RANGE
22–26°C (72–79°F)

COMMUNITY
excellent

ADULT SIZE
female 12cm (4.75in.)
male 12cm (4.75in.)

DIET
all foods

EASE OF KEEPING
10/10

pH RANGE
6.8–7.5

Although of dark coloration, the marbled patterning on this fish makes it very striking. It is an active fish, constantly on the move, searching for food, with its long, slim and tubular shape making it a fast swimmer. It has small, sharp spines, so try to avoid them when catching the fish. Breeding in captivity has not yet been recorded, and sexing them is very difficult.

Feeding

These loaches like a varied diet of flake foods. They readily take live food, such as bloodworms, tubifex and glassworms, as well as live daphnia, once or twice a week.

Khuli loach / *Acanthopthalmus khulii*

The markings on the fish that are sold as Khuli loaches are exceptionally variable. Some have a band all the way round the body; others have bands extending only half way; still more just have a line of large dots across the centre of the back, running the full length of the body. They have an eel-like appearance and are so quick that it is almost impossible to catch them in a planted tank. Khuli loaches swim very fast from one area to another and then settle down for a while. Generally, they swim close to the bottom of the aquarium and make excellent scavengers, eating any food missed by the other inhabitants. They are known to be group-spawners, and the females are plumper than the males, but this is about all that is known of their breeding habits.

Feeding

They like a varied diet of flake foods. They readily take live food, such as bloodworms, tubifex and glassworms, as well as live daphnia, once or twice a week.

ORIGIN
Asia

TEMPERATURE RANGE
22–26°C (72–79°F)

COMMUNITY
excellent

ADULT SIZE
female 10cm (4in.)
male 10cm (4in.)

DIET
all foods

EASE OF KEEPING
10/10

pH RANGE
6.8–7.5

Ringed loach / *Lepidocephalus scatarugu*

ORIGIN
India

TEMPERATURE RANGE
22–26°C (72–79°F)

COMMUNITY
excellent

ADULT SIZE
female 10cm (4in.)
male 10cm (4in.)

DIET
all foods

EASE OF KEEPING
10/10

pH RANGE
6.8–7.5

These fish are rarely available, although they are not especially expensive. They sit at the front of the aquarium, seemingly watching you watching them. They are a community fish, and are no problem when with other fish. Normally sold when they are about 3 to 4cm (1.2 to1.6in.) long, they settle into an aquarium very quickly and feed without any problem. Once they are settled, the rings around the body will become quite dark, and their full pattern begins to show. There is no information at this time as to the breeding or the sexing of these fish.

Feeding

They like a varied diet of flake foods. They readily take live food, such as bloodworms, tubifex and glassworms, as well as live daphnia, once or twice a week.

Bengal loach / *Botia dario*

When this fish was first named, it was decided that its pattern resembled that of the Bengal tiger, which is why it was called the Bengal loach. It has quite wide, sandy-yellow and black bars, and the pattern and colour become much stronger as the fish matures. They tend to be slightly deeper in the body than many other loaches, and this makes them slightly fatter. Although not shy fish, they like to have somewhere to hide away. Very little is known about the breeding or sexing of this fish.

Feeding

They like a varied diet of flake foods. They readily take live food, such as bloodworms, tubifex and glassworms, as well as live daphnia, once or twice a week.

ORIGIN
India

TEMPERATURE RANGE
22–26°C (72–79°F)

COMMUNITY
excellent

ADULT SIZE
female 12cm (4.75in.)
male 12cm (4.75in.)

DIET
all foods

EASE OF KEEPING
10/10

pH RANGE
6.8–7.5

Crossbanded loach / *Botia striata*

ORIGIN
Asia/India

TEMPERATURE RANGE
22–26°C (72–79°F)

COMMUNITY
excellent

ADULT SIZE
female 10cm (4in.)
male 10cm (4in.)

DIET
all foods

EASE OF KEEPING
10/10

pH RANGE
6.8–7.5

This is a very strongly patterned fish, even as a youngster. It has a multitude of thin and thick bands around most of the body, generally chocolate brown in colour. The belly is pale cream and without a pattern. It is a very active fish, and three or four in a tank will normally swim together. It is advisable to provide night-time hiding places. White spot can be a minor problem when they are first moved, but this is easily treated and cured. Little is known about the breeding habits of this fish, except that the female becomes slightly plumper in the belly.

Feeding

They like a varied diet of flake foods. They readily take live food, such as bloodworms, tubifex and glassworms, as well as live daphnia, once or twice a week.

Red rainbowfish / *Glossolepis incissus*

Males are vivid red and females are a plainer, silver colour, so sexing this species is not a problem. As they grow and mature, the body of the male becomes very deep, with a noticeably high back. The female tends to be slightly more elongated. They both grow to a good size, but even when fully grown, are excellent community fish. Breeding is easy, as they are egg-scatterers, but they have to be removed from the eggs once they have finished spawning.

ORIGIN
Papua New Guinea

TEMPERATURE RANGE
22–26°C (72–79°F)

COMMUNITY
excellent

ADULT SIZE
female 13cm (5in.)
male 13cm (5in.)

DIET
all foods

EASE OF KEEPING
10/10

pH RANGE
6.8–7.5

Feeding

They like a varied diet of flake foods. They readily take live food, such as bloodworms, tubifex and glassworms, as well as live daphnia, once or twice a week.

Banded rainbowfish / *Melanotaenia trifasciata*

There are a number of colour and locality varieties of this fish. Based around a red fin colour, the main differences are differing shades and intensities of red. The various varieties are named after the river system in which they are found, for example, Giddy-river variety or Goyder-river variety. The originate in slow-flowing river tributaries, with plenty of overhanging shrubbery. It is advisable to place some large plants in the aquarium for them to hide in. They tend to be rather sedate swimmers, only moving fast when necessary.

Feeding

They like a varied diet of flake foods. They readily take live food, such as bloodworms, tubifex and glassworms, as well as live daphnia, once or twice a week.

ORIGIN
Australia

TEMPERATURE RANGE
22–26°C (72–79°F)

COMMUNITY
excellent

ADULT SIZE
female 12cm (4.75in.)
male 12cm (4.75in.)

DIET
all foods

EASE OF KEEPING
10/10

pH RANGE
6.8–7.5

Neon rainbowfish / *Melanotaenia praecox*

ORIGIN
Papua New Guinea/ Irian Jaya

TEMPERATURE RANGE
22–26°C (72–79°F)

COMMUNITY
excellent

ADULT SIZE
female 5cm (2in.)
male 7cm (2.75in.)

DIET
all foods

EASE OF KEEPING
10/10

pH RANGE
6.8–7.5

This fish was only discovered in about 1994, and since then has become one of the most popular rainbow fish. The male has a beautiful, metallic-blue body, with bright-red edges to the fins, and under aquarium lights the blue body reflects many different shades of blue. They are peaceful, easy to keep and simple to breed. In a lightly planted tank, they will lay between six and eight eggs a day, which they attach to a plant and then leave to hatch, with no parental care. The eggs are small and hard to see; when hatched out, the fry are like tiny hairs.

Feeding

They like a varied diet of flake foods. They readily take live food, such as bloodworms, tubifex and glassworms, as well as live daphnia, once or twice a week.

Black-lined rainbowfish / *Melanotaenia maccullochi*

This rainbowfish has long been popular within the aquatic hobby. When it is settled and happy, it swims with its fins fully erect. The body has a greenish-gold base colour, with a number of black lines along the length of the body, and fins tinged with red, which becomes stronger as the fish matures. This fish is extremely peaceful and, like the other rainbowfish shown in this book, has one extra dorsal fin more than most other fish. The females are not as strongly coloured as the males.

Feeding

They like a varied diet of flake foods. They readily take live food, such as bloodworms, tubifex and glassworms, as well as live daphnia, once or twice a week.

ORIGIN
Australia

TEMPERATURE RANGE
22–26°C (72–79°F)

COMMUNITY
excellent

ADULT SIZE
female 9cm (3.5in.)
male 9cm (3.5in.)

DIET
all foods

EASE OF KEEPING
10/10

pH RANGE
6.8–7.5

Boesmani rainbowfish / *Melanotaenia boesmani*

ORIGIN
Papua New Guinea

TEMPERATURE RANGE
22–26°C (72–79°F)

COMMUNITY
excellent

ADULT SIZE
female 13cm (5in.)
male 13cm (5in.)

DIET
all foods

EASE OF KEEPING
10/10

pH RANGE
6.8–7.5

The colour of this fish is absolutely stunning: the rear half, including the tail, is a bright, strong, orange-yellow and the front half is slate-grey-blue, with two yellow stripes. The body shape is an elongated oval. They like plenty of tall aquatic plants in the aquarium to sit in. They spawn continuously on the plant leaves for two or three days at a time, but then not again for about another month. If you manage to keep the eggs and raise the fry, you will find them very easy to grow and care for.

Feeding

They like a varied diet of flake foods. They readily take live food, such as bloodworms, tubifex and glassworms, as well as live daphnia, once or twice a week.

Thread-fin rainbowfish / *Iratherina werneri*

This fish is extremely slim, and the finnage on the male is exaggerated. Once this fish is settled and well conditioned, the upper body is covered in a metallic lilac and the lower edge of the body has a red hue to it. The elongated fins are jet black, apart from the tail, which is clear, except for red edging to the upper and lower lobes. When the female is ready to spawn and there is more than one male, they will put on a display by showing off to each other, extending their fins and vying for the attentions of the female.

ORIGIN
Papua New Guinea /Australia

TEMPERATURE RANGE
22–26°C (72–79°F)

COMMUNITY
excellent

ADULT SIZE
female 5cm (2in.)
male 5cm (2in.)

DIET
all foods

EASE OF KEEPING
8/10

pH RANGE
6.8–7.5

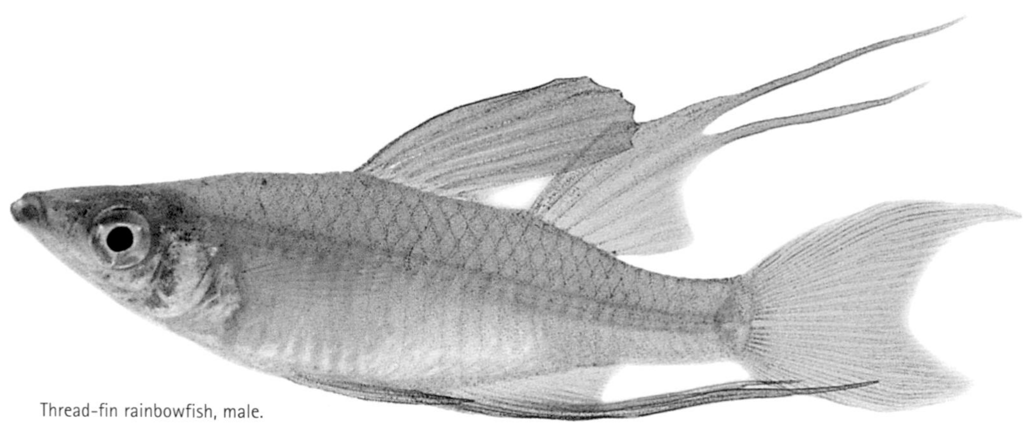

Thread-fin rainbowfish, male.

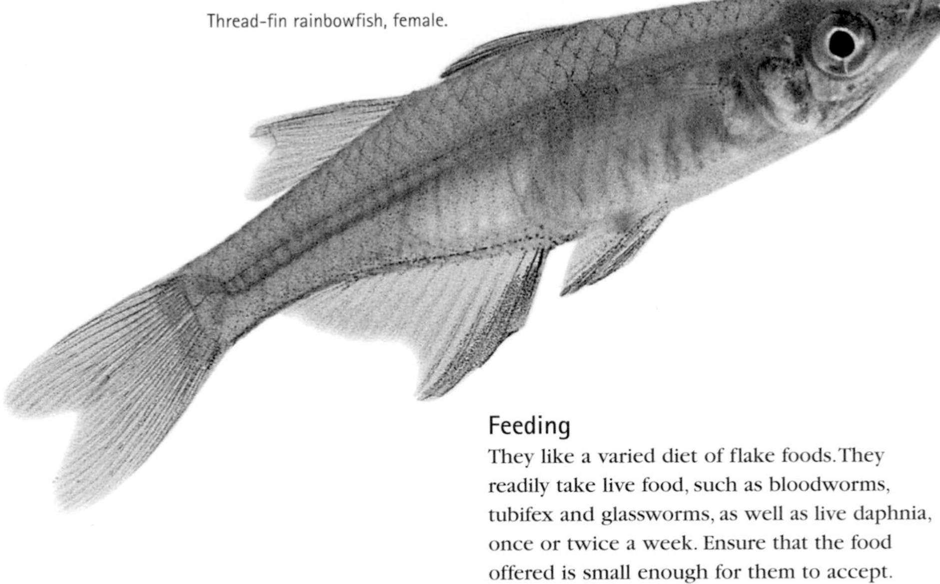

Thread-fin rainbowfish, female.

Feeding

They like a varied diet of flake foods. They readily take live food, such as bloodworms, tubifex and glassworms, as well as live daphnia, once or twice a week. Ensure that the food offered is small enough for them to accept.

Lake Katubu rainbowfish / *Melanotaenia lacustris*

These are highly recommended for the new aquarist, as they are very easy to care for and extremely hardy. They will adapt to new aquarium conditions very quickly. It is advisable to buy these fish in pairs (male and female), as they will swim together and show off to each in their best colours. The upper half of the body is turquoise-blue, fading to white in the lower part of the body. The blue is reflective and seems to change as the fish turn under aquarium lighting. There is also a very dark-blue bar running down the centre of the body. As they become older, the body becomes very deep and the snout of the fish quite pointed. They are typical rainbowfish-spawners, scattering the eggs among the plant.

Feeding

They like a varied diet of flake foods. They readily take live food, such as bloodworms, tubifex and glassworms, as well as live daphnia, once or twice a week.

ORIGIN
Papua New Guinea

TEMPERATURE RANGE
22–26°C (72–79°F)

COMMUNITY
excellent

ADULT SIZE
female 12cm (4.75in.)
male 12cm (4.75in.)

DIET
all foods

EASE OF KEEPING
10/10

pH RANGE
6.8–7.5

Celebes rainbowfish / *Telmatherina ladigesi*

ORIGIN
Indonesia

TEMPERATURE RANGE
22–26°C (72–79°F)

COMMUNITY
excellent

ADULT SIZE
female 7cm (2.75in.)
male 7cm (2.75in.)

DIET
all foods

EASE OF KEEPING
10/10

pH RANGE
6.8–7.5

Celebes rainbowfish were first introduced in 1933, and continue to be popular. When first added to an aquarium, they are liable to hide away until they have established themselves. At first glance, it would seem that they are a type of glassfish, but it is only the rear half of the fish that is transparent. This is a very attractive fish, with a bright-blue line running through the centre of the body in the rear half and a bright-golden-yellow edging to the bottom edge of the body. The fins are black and yellow. As with the other rainbowfishes, they also have twin dorsal fins. The male has longer fins and is stronger in colour than the female.

Feeding

They like a varied diet of flake foods. They readily take live food, such as bloodworms, tubifex and glassworms, as well as live daphnia, once or twice a week.

Harlequin rasbora / *Rasbora heteromorpha*

To get the best from this fish, it is advisable to purchase a shoal of 10 to 12, as they do much better in a group and find security in numbers. Their body shape is unusual, in that the body is quite deep for a small fish and then tapers very quickly up to the tail. The body has a golden hue and a dark-blue-black, triangular wedge shape in the rear half of the body. The fins are mostly red, except for the tail, which is clear. They like plenty of fine-leafed plants in the aquarium for cover. The females are normally deeper than the males and, when ready to spawn, have a stronger gold coloration. This fish can be mistaken for *Rasbora espei,* which is much slimmer. There are now two other colour variations of this fish: black harlequin and gold harlequin.

ORIGIN
Asia

TEMPERATURE RANGE
22–26°C (72–79°F)

COMMUNITY
excellent

ADULT SIZE
female 4.5cm (1.7in.)
male 4.5cm (1.7in.)

DIET
all foods

EASE OF KEEPING
10/10

pH RANGE
6.8–7.5

Harlequin rasbora, standard.

Harlequin rasbora, black.

Feeding

Harlequin rasboras like a varied diet of flake foods. They will readily take live food, such as bloodworms, tubifex and glassworms. Ensure that the food is small enough for them to accept, however.

Scissortail rasbora / *Rasbora trilineata*

This fish is long, slender, agile and a very fast swimmer. A hardy fish, it is easy to maintain and is ideal to place in a new aquarium. The body is semi-transparent, with a black line running from just behind the gill plate to the caudal peduncle. The tail has two black patches, one in each of the tail lobes, and these are bordered on both sides by a yellow patch.

Feeding

Scissortail rasboras like a varied diet of flake foods. They readily take live food, such as bloodworms, tubifex and glassworms, as well as live daphnia, once or twice a week.

ORIGIN
Asia

TEMPERATURE RANGE
22–26°C (72–79°F)

COMMUNITY
excellent

ADULT SIZE
female 10cm (4in.)
male 10cm (4in.)

DIET
all foods

EASE OF KEEPING
10/10

pH RANGE
6.8–7.5

Green-eye rasbora / *Rasbora dorsiocellata*

ORIGIN
Asia

TEMPERATURE RANGE
22–26°C (72–79°F)

COMMUNITY
excellent

ADULT SIZE
female 6.5cm (2.5in.)
male 6.5cm (2.5in.)

DIET
all foods

EASE OF KEEPING
10/10

pH RANGE
6.8–7.5

Unfortunately, this fish is not always readily available in the aquarium shops. It is not a particularly colourful fish, but is very peaceful and extremely hardy. It is an ideal starter fish, especially for the younger aquarist. It swims quite slowly, generally in the middle to lower half of the aquarium. It has a bright-green colour in the lower half of the eye and the dorsal fin has a black-and-white patch on it. The body is quite void of colour. There is no immediate difference between the male and female, although when ready to spawn, the female's belly area is much fuller.

Feeding

These fish like a varied diet of flake foods. They readily take live food, such as bloodworms, tubifex and glassworms, as well as live daphnia, once or twice a week.

Red-tail black shark / *Labeo bicolour*

This group of fish merely resemble sharks: their bodies are a cylindrical shape and the high dorsal fin is also large. The body and fins are normally black and the tail is bright red; there is sometimes a small splash of bright white on the tip of the dorsal fin, which makes them very noticeable fish. When juvenile, this coloration is faded and only comes with maturity. (This is not a sexual difference between male and female.) They are not shy, but are very active, sometimes swimming upside down so that they can graze on the underside of plant leaves. There are few reports of this fish being bred in captivity.

Feeding

These fish like a varied diet of flake foods. They readily take live food, such as bloodworms, tubifex and glassworms, as well as live daphnia, once or twice a week. A small amount of vegetable matter should also be fed to them.

ORIGIN
Asia

TEMPERATURE RANGE
22–26°C (72–79°F)

COMMUNITY
excellent

ADULT SIZE
female 15cm (6in.)
male 15cm (6in.)

DIET
all foods

EASE OF KEEPING
10/10

pH RANGE
6.8–7.5

Red-finned shark / *Labeo erythrurus*

ORIGIN
Asia

TEMPERATURE RANGE
22–26°C (72–79°F)

COMMUNITY
excellent

ADULT SIZE
female 15cm (6in.)
male 15cm (6in.)

DIET
all foods

EASE OF KEEPING
10/10

pH RANGE
6.8–7.5

These fish are very hardy and will accept a wide range of conditions in an aquarium. They have a streamlined shape, which makes them fast swimmers. The body colour is olive-grey and the fins are red. There is a black patch on the caudal peduncle. Like *Labeo bicolor* (see page 185), they are active fish, swimming mainly in the lower half of the aquarium. If there is more than one, they will chase each other from time to time, but normally do no damage. Sexing is very difficult, although the female is slightly more rotund than the male. There is very little information regarding breeding this fish. The albino variety is shown on page 187.

Ruby shark.

Albino ruby shark.

Feeding

These fish like a varied diet of flake foods. They readily take live food, such as bloodworms, tubifex and glassworms, as well as live daphnia, once or twice a week. A small amount of vegetable matter should also be fed.

Bumble-bee goby / *Brachygobius xanthozonus*

ORIGIN
Asia

TEMPERATURE RANGE
22–26°C (72–79°F)

COMMUNITY
excellent

ADULT SIZE
female 5cm (2in.)
male 5cm (2in.)

DIET
all foods

EASE OF KEEPING
8/10

pH RANGE
6.8–7.5

The common name of this fish certainly fits its shape and coloration. They are very small fish, growing to a maximum length of 5cm (2in.), and are slow swimmers that like to sit rocks or leaves for long periods. Larger and faster fish may disturb them, so try to keep them with smaller fish only. It is advisable to have a small shoal in the aquarium, as they will settle down and adjust to the new environment much more quickly. It is thought that the male, when ready to breed, has much stronger coloration, but this is still not proven. There is very little information on the actual breeding of this fish.

Feeding

This fish is very slow in feeding and prefers foods like tubifex, whiteworms, bloodworms and microworms. It will take small quantities of crushed flake food, but does not seem to do very well on it.

Peacock goby / *Tateurndina ocellicauda*

As with many other fish from Papua New Guinea, this is a beautiful species. They have a tubular body shape and are blunt-headed. The body colour of slate blue runs into all of the fins. There are a number of bright-red, broken bars running from top to bottom of the fish, along the full length of the body. There is a red pattern in each of the fins, but each is different. Both of the dorsal fins have a bright-yellow edge, and the anal fin has a red edge. The female is only slightly lighter in colour and a little more full in the belly than the male. They spawn in small caves and the female takes care of the eggs.

Feeding

These fish like a varied diet of flake foods. They readily take live food, such as bloodworms, tubifex and glassworms, as well as live daphnia, once or twice a week.

ORIGIN
Papua New Guinea

TEMPERATURE RANGE
22–26°C (72–79°F)

COMMUNITY
excellent

ADULT SIZE
female 5cm (2in.)
male 5cm (2in.)

DIET
all foods

EASE OF KEEPING
10/10

pH RANGE
6.8–7.5

Glass hatchet fish / *Carnigiella myersi*

ORIGIN
South America

TEMPERATURE RANGE
22–26°C (72–79°F)

COMMUNITY
excellent

ADULT SIZE
female 3cm (1in.)
male 3cm (1in.)

DIET
all foods

EASE OF KEEPING
8/10

pH RANGE
6.8–7.5

These fish have a very peculiar shape: from above, they are appear extremely thin-bodied and also have extremely large pectoral fins. If frightened or disturbed, they will use the power from their pectoral fins to thrust themselves out of the water and seemingly fly, so keep a tight-fitting lid on the aquarium. For most of the time they swim very close to the surface. There is no information on sexual differences or the breeding of this fish.

Feeding

When feeding, the lower jaw of this fish drops, and it can easily take floating foods. This fish eats a varied diet of flake foods, and readily takes live food, such as glassworms or daphnia. A treat of black mosquito larvae is very welcome, and is available freeze-dried.

Silver hatchet fish / *Thoracocharax stellatus*

This is a larger hatchet fish than many, which is also thin-bodied. It never fattens out and has extremely large pectoral fins, which it uses to thrust itself out of the water when frightened. It can 'fly' up to about 3m (9ft) to escape a predator, so keep a tight-fitting lid on the aquarium. It swims close to the surface most of the time. There is no information on sexual differences or the breeding of this fish.

Feeding

When feeding, the lower jaw of this fish drops and it can easily take floating foods. This fish eats a varied diet of flake foods and readily takes live food, such as glassworms or daphnia. A treat of black mosquito larvae is very welcome, and is available freeze-dried.

ORIGIN
South America

TEMPERATURE RANGE
22–26°C (72–79°F)

COMMUNITY
excellent

ADULT SIZE
female 8cm (3in.)
male 8cm (3in.)

DIET
all foods

EASE OF KEEPING
10/10

pH RANGE
6.8–7.5

Marbled hatchet fish / *Carnigiella strigata*

ORIGIN
South America

TEMPERATURE RANGE
22–26°C (72–79°F)

COMMUNITY
excellent

ADULT SIZE
female 5cm (2in.)
male 5cm (2in.)

DIET
all foods

EASE OF KEEPING
10/10

pH RANGE
6.8–7.5

With a black, marbled pattern on its body, this fish is easy to identify. Like the other hatchets shown, it is thin-bodied, with large pectoral fins. It swims close to the surface, but there is virtually no information on breeding.

Feeding

When feeding, the lower jaw of this fish drops and it can easily take floating foods. This fish eats a varied diet of flake foods and readily takes live food, such as glassworms or daphnia. A treat of black mosquito larvae is very welcome, and is available freeze-dried.

Long-nose elephant fish / *Gnathonemus petersii*

The long-nose elephant fish is a good community fish, but has some special requirements. A sand substrate is preferred because they use their extended snouts to push into the sand looking for food and cannot do this in a gravel substrate. Although they are quite a large fish, they can be shy, so plenty of cover is required. It would be better to place only smaller, quieter fish with them. With a dark, chocolate-brown body coloration and two pinkish bars towards the rear of the body, they stand out against a sand substrate, which must be silica sand, not builder's sand. There is very little known about sexual differences or the breeding of this fish.

Feeding

Flake food is not very good for these fish, so feed tubifex or bloodworms, which burrow into the substrate, allowing the fish to feed naturally.

ORIGIN
Africa

TEMPERATURE RANGE
22–26°C (72–79°F)

COMMUNITY
excellent

ADULT SIZE
female 25cm (9in.)
male 25cm (9in.)

DIET
all foods

EASE OF KEEPING
8/10

pH RANGE
6.8–7.5

Flying fox / *Epalzeoryhnchus kallopterus*

ORIGIN
Asia

TEMPERATURE RANGE
22–26°C (72–79°F)

COMMUNITY
excellent

ADULT SIZE
female 15cm (6in.)
male 15cm (6in.)

DIET
all foods

EASE OF KEEPING
10/10

pH RANGE
6.8–7.5

The flying fox has become extremely popular with aquarists, although it is only available at certain times of the year. Hardy, and easy to keep, they are fast fish, although they do settle down and sit very high on their pectoral fins, looking out of the aquarium. They are attractive fish, with a black-and-gold stripe running from the tip of the snout right through to the central end of the tail. The very top of the back also has a black stripe running the length of the body. The fins have black smudges in them. Two or more in an aquarium will chase each other quite harmlessly.

Feeding

These fish like a varied diet of flake foods. They readily take live food, such as bloodworms, tubifex and glassworms, as well as live daphnia, once or twice a week.

Golden-apple snail / *Ampullaria cuprina*

Snails are not usually popular in aquaria, but these graze over the tank, cleaning off algae and creating infusoria from their waste matter, which is ideal food for recently hatched fry. They require slightly hard water to maintain the strength of their shell, which is a bright-yellow-gold coloration. Breeding is easy: they climb above the level of the water and lay pink eggs on top of each other on the aquarium glass until they resemble a raspberry. As long as the eggs do not become too dry, they will hatch in about three weeks and then make their way back down into the aquarium. These snails do not like bad water quality and regular water changes should be made.

Feeding

Scalded lettuce leaves are their main diet.

ORIGIN
Asia

TEMPERATURE RANGE
22–26°C (72–79°F)

COMMUNITY
excellent

ADULT SIZE
female 7cm (2.75in.)
male 7cm (2.75in.)

DIET
all foods

EASE OF KEEPING
10/10

pH RANGE
6.8–7.5

Guppy / millions fish / *Poecilia reticulatus*

ORIGIN
Central America

TEMPERATURE RANGE
22–26°C (72–79°F)

COMMUNITY
excellent

ADULT SIZE
female 5cm (2in.)
male 7cm (2.75in.)

DIET
all foods

EASE OF KEEPING
10/10

pH RANGE
6.8–7.5

Guppies are one of the best-known tropical fish and appear with many different colour forms and tail shapes. They are generally very hardy and accept a wide range of conditions in an aquarium, but because their tail is so delicate, do not add them to a new aquarium during the first six weeks.

Sexing is easy because the male has a flowing tail and highly coloured, well-patterned body, while females are drab by comparison. Males constantly chase the females, so if you wanted to breed selectively, mix the fish with care. Most commercially bred fish are produced in Far Eastern countries, such as Singapore, Malaysia, Thailand and Sri Lanka.

Among the different types of guppy available are roundtail, lyretail, top sword, bottom sword, double sword, pintail, veiltail and many others. Some are only available from specialist breeders, but there are a number of specialist guppy or live-bearer societies that are the source of a great deal of information and specialist knowledge.

When pregnant, females have a black blotch at the rear of the belly area, known as a gravid spot. Under normal conditions, the gestation period is between 28 and 35 days. Being a live-bearing fish, as the fry are released they are free-swimming immediately and swim for cover and try to stay well hidden for a while. They produce about 70 or 80 young at one time, and can bear more than one brood of fry from one fertilisation, but numbers of fry will reduce.

Feeding

These fish like a varied diet of flake foods. They readily take live food, such as bloodworms, tubifex and glassworms, as well as live daphnia, once or twice a week.

Mosaic guppy, male.

Blonde-red tuxedo guppy, male.

Dragonhead tuxedo guppy, male.

Rainbow guppy, male.

Gold guppy, male.

Neon-blue guppy, male.

Neon-red guppy, male.

Flame guppy, male.

Green snakeskin guppy, male.

Metallic-red-tail guppy, male.

Guppy, female, colour variation .

Guppy, female, colour variation .

Guppy, female, colour variation .

Guppy, female, colour variation .

Guppy, female, colour variation .

Platy / *Xiphophorus maculatus*

Platys are another well-known species of tropical fish worldwide. They are very hardy and tolerate a wide range of conditions in an aquarium, making them ideal to place in a new aquarium.

The male has a modified anal fin that is called a gonopodium. Males constantly chase the females, so if you wanted to breed selectively, mix the fish with care.

Some platys are only available from specialist breeders and may be difficult to acquire. Unfortunately, there is not enough demand for them to be bred commercially, although a specialist live-bearer society will be a source of valuable information. You can normally find the contacts for these groups in aquatic magazines.

Like guppies, pregnant females display a gravid spot on the rear of the belly and gestation is between 28 and 35 days. Between 70 and 80 free-swimming fry are released and swim straight for plant cover, as there is no parental care for the young fish.

Feeding

These fish like a varied diet of flake foods. They readily take live food, such as bloodworms, tubifex and glassworms, as well as live daphnia, once or twice a week.

ORIGIN
Central America

TEMPERATURE RANGE
22–26°C (72–79°F)

COMMUNITY
excellent

ADULT SIZE
female 5cm (2in.)
male 7cm (2.75in.)

DIET
all foods

EASE OF KEEPING
10/10

pH RANGE
6.8–7.5

Spotted platy, male.

Spotted platy, female.

Gold comet platy, female.

Gold comet platy, male.

Red wag platy, male.

Red wag platy, female.

Black platy, male.

Black platy, female.

Blue coral platy, male.

Blue coral platy, female.

Leopard platy, male.

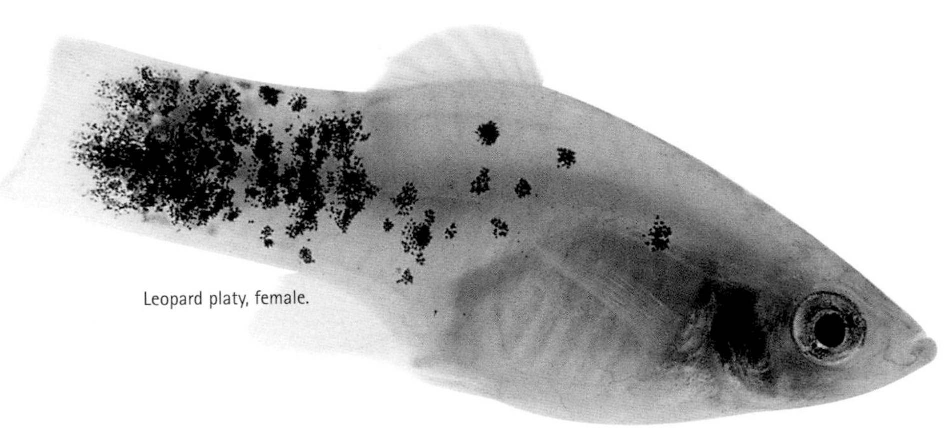
Leopard platy, female.

Blue Mickey Mouse platy, male.

Blue Mickey Mouse platy, female.

Gold wag platy, male.

Gold wag platy, female.

Swordtail / *Xiphophorus helleri*

The swordtail is a popular tropical fish, but it is unusual to see all of its varieties available. They are a very hardy fish and accept a wide range of conditions in an aquarium, so they are ideal additions to a newly established aquarium.

The male has a modified anal fin called a gonopodium, and the lower lobe of the tail has an extension which can be as long as half the length of the body, which is what gives the fish its common name. The males chase the females in an attempt to mate with them, so if you wanted to breed selectively, mix the fish with care. Most of these fish are bred in the Far East, and other types of swordtail are only available from specialist breeders. Unfortunately, there is not enough demand for them to be bred commercially.

Like the guppy and the platy, pregnant females show a gravid spot, a black blotch at the rear of the belly area. Swordtails are live-bearing fish, and after a gestation of between 28 and 35 days free-swimming fry are born, which swim straight for cover among the plants. The adult female can have up to about 70 or 80 young at one time, but there is no parental care for the young fish. The females are able to have more than one brood of fry from one fertilisation, but the numbers of fry are reduced.

Feeding

These fish like a varied diet of flake foods. They readily take live food, such as bloodworms, tubifex and glassworms, as well as live daphnia, once or twice a week.

ORIGIN
Central America

TEMPERATURE RANGE
22–26°C (72–79°F)

COMMUNITY
excellent

ADULT SIZE
female 10cm (4in.)
male 11cm (4.3in.)

DIET
all foods

EASE OF KEEPING
10/10

PH RANGE
6.8–7.5

Black swordtail, male.

Black swordtail, female.

Red-and-white swordtail, male.

Red-and-white swordtail, female.

Tuxedo swordtail, male.

Tuxedo swordtail, female.

Red-streak swordtail, male.

Red-streak swordtail, female.

Red wag swordtail, male.

Red wag swordtail, female.

Red swordtail, male.

Red swordtail, female.

Lyretail swordtail, male.

Lyretail swordtail, female.

Green swordtail, male.

Green swordtail, female.

Molly / *Poecilia velifera*

Mollies are a popular tropical fish and are available in a wide variety of colours. They are hardy fish and accept a wide range of conditions in an aquarium, but prefer a small amount of aquarium salt in the tank (1 teaspoon per gallon or 4.5 litres). They are prone to 'shimmying', which is caused by shock or a chilling of the fish. A salt bath in 1 gallon (4.5 litres) of water from the aquarium and 4 teaspoons of thoroughly dissolved salt for 20 minutes once a day for three days should solve this problem. They also suffer occasionally from oodinium, which is often mistaken for white spot. A combination of salt and a proprietary medication will very easily cure this problem.

The male has a modified anal fin known as a gonopodium and constantly chases the females, so if you wanted to breed selectively, mix the fish with care. Other types of mollies are only available from specialist breeders; unfortunately there is not enough demand for them to be bred commercially. Specialist live-bearer societies are an excellent source of information.

Mollies are live-bearing fish, and after a gestation of between 28 and 35 days free-swimming fry are born, which swim straight for cover among the plants. The adult female can have up to about 50 or 60 young at one time, but there is no parental care for the young fish. The females are able to have more than one brood of fry from one fertilisation, but the numbers of fry are reduced.

ORIGIN
Central America

TEMPERATURE RANGE
22–26°C (72–79°F)

COMMUNITY
excellent

ADULT SIZE
female 10cm (4in.)
male 11cm (4.2in.)

DIET
mainly vegetables

EASE OF KEEPING
10/10

pH RANGE
6.8–7.5

Gold marble sailfin molly, male.

Feeding

Mollies like a varied diet of flake foods. They readily take live food, such as bloodworms, tubifex and glassworms, as well as live daphnia, once or twice a week.

Gold marble sailfin molly, female.

Green sailfin

Green sailfin molly, female.

Black sailfin molly, male.

Black sailfin molly, female.

Silver molly, female.

Sunshine molly, male.

Silver sailfin molly, male.

Copper spotted molly, female.

Polkadot sailfin molly, male.

Fantail / *Carassius* sp.

ORIGIN
China

TEMPERATURE RANGE
20–25°C (68–77°F)

COMMUNITY
excellent

ADULT SIZE
female 15cm (6in.)
male 15cm (6in.)

DIET
all foods

EASE OF KEEPING
10/10

pH RANGE
6.8–7.5

Most fantails or oranda varieties originate in China, and some have become more than just a colour variety. 'Celestials', for example, have their eyes set very high in the head, looking upwards towards the sky; 'bubble-eyes' have large sacs that look like big bubbles attached to the head; and 'lionheads' have what is called a 'bramble' on the top of the head. Chinese breeders have been selectively breeding these fish and creating new varieties for hundreds of years, and just about every type of fantail in every colour combination is available at some time throughout the year.

Although fantails are regarded as cold-water fish, and many people do put them in their ponds, this is not really advisable. As autumn arrives, the water becomes much too cold and they have to be placed in a large aquarium for the winter. There are always exceptions, and a few may survive winter conditions in a pond, but this is not the norm. These fish will thrive in an aquarium with a maximum temperature of about 26°C (79°F). They mix well with tropical community fish that also prefer a slightly lower temperature, but ensure that the other fish will not harass them or nibble their large, flowing fins. If the temperature in the aquarium is too high, the metabolism of the fish speeds up and their life span will be reduced. Regular water changes are needed, because fantails produce a large volume of waste matter, which may pollute the water, especially if the filtration system is not efficient. They also require a higher oxygen level than conventional tropical fish.

Red fantail.

Red-and-black fantail.

Feeding

These fish like a varied diet of flake foods. They readily take live food, such as bloodworms, tubifex and glassworms, as well as live daphnia, once or twice a week.

Red-and-white fantail.

Calico fantail.

Blackmoor fantail.

Marliers Julie / *Julidochromis marlieri*

In the *Julidochromis* genus, there are only five accepted species. There are a number of subspecies, which are location varieties where the pattern or coloration is slightly different from the original species. These fish require an aquarium set-up with plenty of rocks and large, round stones so that they create their own territories and breeding areas as they are cave-spawners. When a compatible pairing has been made, they will clean out an area and then go through the mating process. When the eggs hatch, the parents take it in turn to care for the youngsters and will defend them quite vigorously. They are community fish, but are best placed either in an aquarium with larger fish or one that has been specifically set up as an African-lake biotope. They are hardy fish and very easy to maintain.

Feeding

These fish are quite happy to take a good, varied diet of different high-quality flake foods. They will readily take live food. such as bloodworms, tubifex and glassworms. Feed live daphnia once or twice a week.

ORIGIN
Africa (Lake Tanganyika)

TEMPERATURE RANGE
22–26°C (72–79°F)

COMMUNITY
excellent

ADULT SIZE
female 15cm (6in.)
male 15cm (6in.)

DIET
all foods

EASE OF KEEPING
10/10

PH RANGE
6.8–7.5

Masked Julie / *Julidochromis transcriptus kissi*

ORIGIN
Africa (Lake Tanganyika)

TEMPERATURE RANGE
22–26°C (72–79°F)

COMMUNITY
excellent (with larger fish only)

ADULT SIZE
female 9cm (3.5in.)
male 9cm (3.5in.)

DIET
all foods

EASE OF KEEPING
10/10

pH RANGE
6.8–7.5

This subspecies has a slightly heavier pattern than the original *Julidochromis transcriptus* and is a location variety from the Kissi area at the northern end of Lake Tanganyika. Cave-spawners, these fish require an aquarium with plenty of rocks and large, round stones to create their own territories and breeding areas. A breeding pair will clean out an area and then go through the mating process. When the eggs hatch, both parents care for the youngsters and will defend them quite vigorously. They are community fish, but are best placed either in an aquarium with larger fish or in one that has been specifically set up as an African-lake biotope. They are hardy fish and very easy to maintain.

Feeding

These fish are quite happy to take a good, varied diet of different high-quality flake foods. They will readily take live food, such as bloodworms, tubifex and glassworms. Feed live daphnia once or twice a week.

Lyretail lamprologus / *Neolamprologus brichardi*

Formerly known as ***Lamprologus brichardi***, this fish has been popular with aquarists for many years. They like an aquarium containing plenty of rocks and large, round stones as they are cave-spawners. Breeding pairs take care of the youngsters together and will defend them quite vigorously. Community fish, they should only be placed either in an aquarium with larger fish or one that has been set-up as an African-lake biotope. This species will quite happily live in a shoal and is hardy and easy to maintain.

Feeding

These fish like a good, varied diet of different flake foods. They will readily take live food, such as bloodworms, tubifex and glassworms. Feed live daphnia once or twice a week.

ORIGIN
Africa (Lake Tanganyika)

TEMPERATURE RANGE
22–26°C (72–79°F)

COMMUNITY
excellent

ADULT SIZE
female 10cm (4in.)
male 10cm (4in.)

DIET
all foods

EASE OF KEEPING
10/10

pH RANGE
6.8–7.5

Orange-spot lamprologus / *N. oscellatus orange*

ORIGIN
Africa
(Lake Tanganyika)

TEMPERATURE RANGE
22–26°C (72–79°F)

COMMUNITY
excellent (with larger fish only)

ADULT SIZE
female 10cm (4in.)
male 10cm (4in.)

DIET
all foods

EASE OF KEEPING
10/10

pH RANGE
6.8–7.5

This is a colour variety of N.oscellatus that has only recently become available to the hobbyist. They prefer an aquarium set-up with plenty of rocks and large round stones, but are actually shell-dwellers. A breeding pair will search for and clean out a large enough shell for themselves and then go through the mating process. When the eggs hatch, both parents care for the youngsters. Community fish, they should only be placed in an aquarium with larger fish or one that is set-up as an African lake biotope. This species will quite happily live in a shoal, and is hardy and easy to maintain.

Feeding

These fish like a good varied diet of different flake foods. They will readily take live food such as bloodworm, tubifex and glassworm. Feed live daphnia once or twice a week.

Brevis shell-dweller / *Neolamprologus brevis*

These are very brightly patterned fish, with white stripes down the body and yellow edging to the dorsal fin and tail. There is also a black edging to the anal fin when they are ready to breed. They prefer an aquarium with plenty of rocks and large, round stones, but are actually shell-dwellers. A breeding pair will search for, and clean out, a large enough shell for themselves and then go through the mating process. When the eggs hatch, both parents care for the youngsters and will defend them quite vigorously. Community fish, they should only be placed in an aquarium with larger fish or one that has been set up as an African-lake biotope. This species will quite happily live in a shoal and is hardy and easy to maintain.

Feeding

These fish like a good, varied diet of different flake foods. They will readily take live food, such as bloodworms, tubifex and glassworms. Feed live daphnia once or twice a week.

ORIGIN
Africa (Lake Tanganyika)

TEMPERATURE RANGE
22–26°C (72–79°F)

COMMUNITY
very good

ADULT SIZE
female 6cm (2.4in.)
male 6cm (2.4in.)

DIET
all foods

EASE OF KEEPING
10/10

pH RANGE
6.8–7.5

White-spotted cichlid / *Tropheus duboisi*

ORIGIN
Africa
(Lake Tanganyika)

TEMPERATURE RANGE
22–26°C (72–79°F)

COMMUNITY
excellent (with larger fish only)

ADULT SIZE
female 12cm (4.75in.)
male 12cm (4.75in.)

DIET
all foods

EASE OF KEEPING
10/10

pH RANGE
6.8–7.5

As a juvenile, *Tropheus duboisi* has a black body with very bright-white spots; as it grows, the pattern is transformed. As some of the spots begin to fade and disappear, a patchy, wide band appears in their place; eventually the spots disappear completely, leaving a bright, wide band in their place on a black body coloration. They prefer a rocky set-up in the aquarium to provide security and breeding areas. They will care for their young for the first couple of weeks, defending them against potential predators. This is a very hardy fish that is easy to maintain.

Feeding

These fish like a good, varied diet of different flake foods. They will readily take live food, such as bloodworms, tubifex and glassworms. Feed live daphnia once or twice a week.

Frontosa cichlid / *Cyphotilapia frontosa*

The Frontosa cichlid grows to quite a large size and requires an aquarium at least 2m (6ft) long and 60cm (24in.) high. They swim in a very leisurely, serene way, apparently just drifting around the aquarium. The adult male has a cranial hump at the front of the head, which is very distinct, and the female has a smaller one. They spawn only about 50 eggs, but take great care of them by fanning and cleaning them until they have hatched out. The fry are kept in a tight group until they are big enough to feed and fend for themselves. Given patience, they can become tame enough to feed directly from your hand.

Feeding

These fish like a varied diet of different flake foods. They will readily take live food, such as bloodworms, tubifex and glassworms. Feed live daphnia once or twice a week.

ORIGIN
Africa (Lake Tanganyika)

TEMPERATURE RANGE
22–26°C (72–79°F)

COMMUNITY
excellent (preferably with own species)

ADULT SIZE
female 35cm (13.75in.)
male 35cm (13.75in.)

DIET
all foods

EASE OF KEEPING
10/10

pH RANGE
6.8–7.5

Pseudotropheus red / blue / *Pseudotropheus* red/blue sp.

ORIGIN
Africa (Lake Malawi)

TEMPERATURE RANGE
22–26°C (72–79°F)

COMMUNITY
good with larger fish only

ADULT SIZE
female 15cm (6in.)
male 15cm (6in.)

DIET
all foods

EASE OF KEEPING
10/10

pH RANGE
6.8–7.5

These are ideal starter fish if set up an aquarium specifically for African cichlids. They are exceptionally hardy fish and will accept quite variable conditions in an aquarium. They prefer a rocky set-up to give them security and breeding areas. They are easy to sex, as the male is one colour and the female the other. They are also quite easy to breed and, like most lake cichlids, will care for their young for the first couple of weeks, defending them against predators.

Pseudotropheus red.

Feeding

These fish like a varied diet of different flake foods. They will readily take live food, such as bloodworms, tubifex and glassworms. Feed live daphnia once or twice a week.

Pseudotropheus blue.

Firemouth / *Thorichthys meeki*

ORIGIN
South America

TEMPERATURE RANGE
22–26°C (72–79°F)

COMMUNITY
very good with fish of same size

ADULT SIZE
female 15cm (6in.)
male 15cm (6in.)

DIET
all foods

EASE OF KEEPING
10/10

pH RANGE
6.8–7.5

These would be an ideal starter fish for an aquarium of larger community fish. They are exceptionally hardy fish, will accept variable conditions in an aquarium and are very easy to care for. When in breeding condition, the male has a vivid-red throat area and shows off to the other males in the aquarium by flaring his large gill-plate covers, vying for the attention of the female. Once a pair has spawned and laid up to 200 eggs on a flat stone, they will care for them by fanning them. They will defend the eggs, and the young once they have hatched, against potential predators.

Feeding

These fish like a varied diet of different flake foods. They will readily take live food, such as bloodworms, tubifex and glassworms. Feed live daphnia once or twice a week.

Daemonsoni cichlid / *Pseudotropheus daemonsoni*

This is an ideal starter fish for an aquarium of African cichlids. They are exceptionally hardy fish, will also accept variable conditions in an aquarium and are very easy to care for. They can be aggressive, but if placed in an aquarium with plenty of rockwork and large stones and a large stock of other African cichlids, they will be content to set up their own territory, merely chasing the other fish harmlessly. As with many of the other African-lake cichlids, both males and females are brightly or strongly coloured. The males will pair with more than one female at the same time. The eggs are laid in caves, so you will be unable to see them until they hatch.

Feeding

These fish like a varied diet of different flake foods. They will readily take live food, such as bloodworms, tubifex and glassworms, and finely grated beef heart is also popular.

ORIGIN
Africa

TEMPERATURE RANGE
22–26°C (72–79°F)

COMMUNITY
very good (but needs to be an African tank)

ADULT SIZE
female 15cm (6in.)
male 15cm (6in.)

DIET
all foods

EASE OF KEEPING
10/10

pH RANGE
6.8–7.5

Kribensis / *Pelvicachromis pulcher*

ORIGIN
Africa

TEMPERATURE RANGE
22–26°C (72–79°F)

COMMUNITY
very good with larger fish

ADULT SIZE
female 7cm (2.75in.)
male 10cm (4in.)

DIET
all foods

EASE OF KEEPING
10/10

PH RANGE
6.8–7.5

The kribensis will live quite happily with other community fish, such as large platys, swordtails and corydoras, but certainly not with the smaller tetras or danios, for example. Any community fish that is larger than 6 to 7cm (2.4 to 2.75in.) would probably be fine. The male is the much larger of the pair and also has longer fins. The female is a much stubbier fish, which, when ready to breed, has a dark-red belly area. They are strong fish and very easy to care for. Commercially bred examples are most widely available, but subspecies are imported from the wild with different patterns and coloration.

Feeding

These fish prefer a varied diet of different flake foods. They will readily take live food, such as bloodworms, tubifex and glassworms, and also like finely grated beef heart.

Sheepshead acara / *Laetocara curviceps*

As with the kribensis, this fish will live quite happily with other community fish, such as large platys, swordtails, corydoras or any community fish larger than 6 to 7cm (2.4 to 2.75in.). The male is the larger of the pair and also has longer fins. The female is much fuller in the body. These fish lay their eggs on large plant leaves, such as Amazon swords, or large, smooth stones or slate. Typically cichlid, they protect the fry from the other inhabitants and will care for their young like this for about seven to ten days. They are strong fish and very easy to care for.

Feeding

These fish prefer a varied diet of different flake foods. They will readily take live food, such as bloodworms, tubifex and glassworms, and also like finely grated beef heart.

ORIGIN
South America

TEMPERATURE RANGE
22–26°C (72–79°F)

COMMUNITY
very good with larger fish

ADULT SIZE
female 8cm (3.2in.)
male 10cm (4in.)

DIET
all foods

EASE OF KEEPING
10/10

pH RANGE
6.8–7.5

Spotted headstander / *Chilodus punctatus*

ORIGIN
South America

TEMPERATURE RANGE
22–26°C (72–79°F)

COMMUNITY
excellent with quiet fish

ADULT SIZE
female 10cm (4in.)
male 10cm (4in.)

DIET
all foods

EASE OF KEEPING
8/10

pH RANGE
6.8–7.5

This headstander always swims nose down at a very steep angle. They are very timid fish and need to be kept with other slow-swimming, non-aggressive fish. They like the security of a well-planted aquarium and will normally take four to five days to settle into a new environment. They are prone to white spot because of the stress of being moved. Sexing this fish is very difficult. It is known that they scatter their eggs amongst plants, but give no parental care to the eggs.

Feeding

These fish prefer a varied diet of different flake foods. They will readily take live food, such as bloodworms, tubifex and glassworms. They can be slow to feed, so ensure that sufficient food is placed in the aquarium for them without polluting the aquarium.

Piranha / *Serrasalmus nattereri*

This fish generally has a poor reputation as an aggressive fish. They can be exceptionally vicious and predatory fish if not fed enough or correctly. They are most definitely not a fish to place in any community aquarium, as they will probably eat the other inhabitants. They live in very large shoals in the wild, an environment that is hard to recreate. They will chase each other and take bites out of each other's fins, so it is very unusual to see a piranha in excellent condition. In the event that you do want to keep this fish, they are very hardy and not prone to disease. They spawn in groups, scatter the eggs and try to devour them.

ORIGIN
South America

TEMPERATURE RANGE
22–26°C (72–79°F)

COMMUNITY
only with own type

ADULT SIZE
female 25cm (10in.)
male 25cm (10in.)

DIET
all foods

EASE OF KEEPING
10/10

pH RANGE
6.8–7.5

Piranha mouth.

Feeding

These fish prefer a varied diet of different flake foods. They will readily take live food, such as bloodworms, tubifex and glassworms.

Black-ghost knife fish / *Sternarchus albifrons*

This long, slim, black fish likes a well-planted aquarium so that it can just vanish into the shadows amongst the plants and rest. If there are no plants, the fish will rest by lying on the substrate or behind a rock on its side and will appear to be dead. In the wild, this fish senses its food by giving off electrical impulses to locate and find it. They have very small eyes for the size of their body. They are shy fish and do not like major changes in the water quality of an aquarium.

Feeding

These fish will eat a varied diet of different flake foods, but it is not the best for them. Live food, such as bloodworms, tubifex, chopped small earthworms and glassworms are the best type of feed for them.

ORIGIN
South America

TEMPERATURE RANGE
22–26°C (72–79°F)

COMMUNITY
only with larger, non-agressive fish

ADULT SIZE
female 20cm (8in.)
male 20cm (8in.)

DIET
all foods

EASE OF KEEPING
9/10

pH RANGE
6.8–7.5

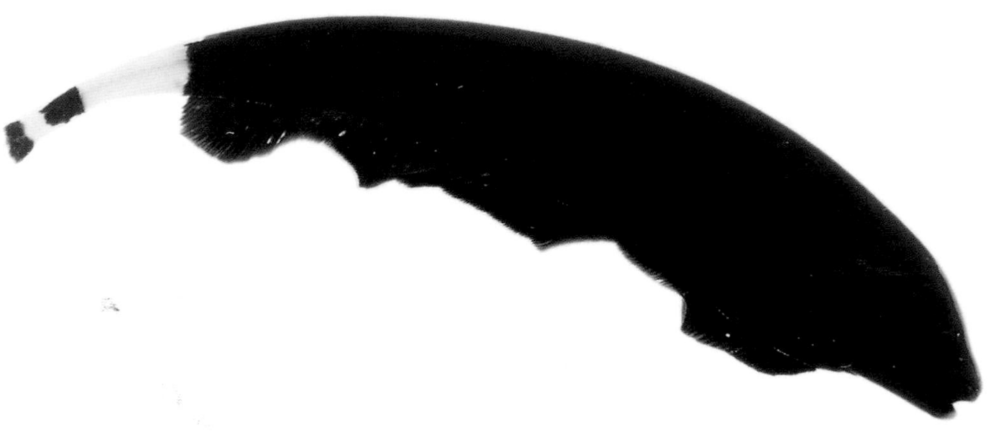

Red scat / *Scatophagus rubrifrons*

ORIGIN
Sri Lanka

TEMPERATURE RANGE
22–26°C (72–79°F)

COMMUNITY
only with larger, non-agressive fish

ADULT SIZE
female 20cm (8in.)
male 20cm (8in.)

DIET
mainly vegetables

EASE OF KEEPING
9/10

pH RANGE
6.8–7.5

Try not to keep this fish in a well-planted aquarium, as it will eat the plants. They grow quite large and their food requirements are substantial. Use bamboo canes and a sand substrate in as large a tank as possible. They also require some salt in the aquarium as they are a brackish-water fish. As they become adult, they will require higher levels of salt. In the wild, they move to, and breed in, fresh water, but return to the brackish tributaries where they originated. High levels of filtration, water quality and oxygen are also needed. They are a schooling fish and are normally very peaceful. There is no information about breeding or sexing this fish.

Feeding

These fish are quite happy to take a varied diet of different, high-quality flake foods. They eat live food, such as bloodworms, tubifex and glassworms, but lettuce is their preferred food.

Green scat / *Scatophagus argus*

Like the red scat (see page 244), the green scat will eat aquarium plants and has substantial food requirements. Use bamboo canes and a sand substrate in as large a tank as possible. They also require some salt in the aquarium as they are a brackish-water fish. As they become adult, they will require higher levels of salt. In the wild, they move to, and breed in, fresh water, but return to the brackish tributaries where they originated. High levels of filtration, water quality and oxygen are also needed. They are a schooling fish and are normally very peaceful. There is no information about breeding or sexing this fish.

Feeding

These fish are quite happy to take a varied diet of different, high-quality flake foods. They eat live food, such as bloodworms, tubifex and glassworms, but lettuce is their preferred food.

ORIGIN
Asia

TEMPERATURE RANGE
22–26°C (72–79°F)

COMMUNITY
only with larger, non-agressive fish

ADULT SIZE
female 20cm (8in.)
male 20cm (8in.)

DIET
mainly vegetables

EASE OF KEEPING
9/10

pH RANGE
6.8–7.5

Teacup stingray / *Potomotrygon sp.*

ORIGIN
South America

TEMPERATURE RANGE
22–26°C (72–79°F)

COMMUNITY
only with larger, non-agressive fish

ADULT SIZE
female 25cm (9in.)
male 25cm (9in.)

DIET
mainly river shrimp

EASE OF KEEPING
8/10

pH RANGE
6.8–7.5

There is still a great deal to learn about stingrays. The teacup stingray is normally available at about 10cm (3.9in.) and, once settled, grows very well. A sand base is vital in the aquarium so that they can bury themselves. They will not tolerate nitrites in the water, so regular water changes are a must, as is a high level of oxygenation and filtration. They will take time to settle into any new aquarium, but are a pleasure to keep. They are active fish, swimming up the sides of the aquarium.

Feeding

These fish will eat a varied diet of different flake foods. Live food, such as bloodworms and tubifex, are accepted, but they prefer regular feeds of river shrimp. They will also accept small pieces of cooked white fish.

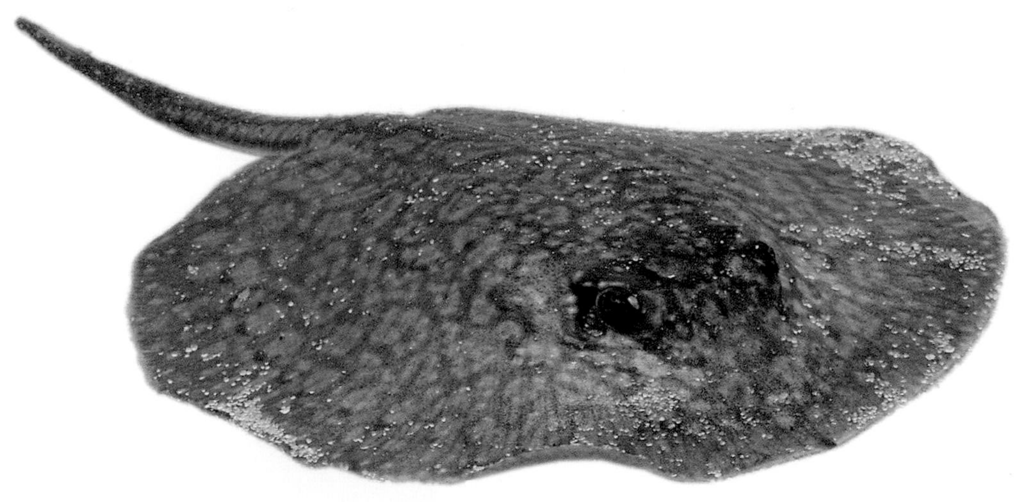

Striped barb / *Barbus fasciatus*

The striped barb grows into a clean-lined, stocky fish. The male is slightly slimmer than the female. They are fast swimmers and are constantly on the move. They need to be kept with larger community fish because they tend to swim 'through' smaller fish rather than around them. When they are very young they have bars down the body, which, as they start to grow, changes to a chequerboard pattern and then to stripes as they become adult. For many years, only juveniles or adults were imported and these were regarded as two different species. They are a typical egg-scatterer, laying their eggs among fine-leafed plants and returning to devour them. They are generally a very hardy and easy fish to keep.

Feeding

These fish prefer a varied diet of different flake foods. Live food, such as bloodworms, tubifex and glassworms, are readily taken.

ORIGIN
Africa

TEMPERATURE RANGE
22–26°C (72–79°F)

COMMUNITY
only with larger fish

ADULT SIZE
female 12cm (4.75in.)
male 12cm (4.75in.)

DIET
all foods

EASE OF KEEPING
9/10

pH RANGE
6.8–7.5

Spanner barb / *Barbus lateristriga*

ORIGIN
Asia

TEMPERATURE RANGE
22–26°C (72–79°F)

COMMUNITY
only with larger fish

ADULT SIZE
female 18cm (7in.)
male 18cm (7in.)

DIET
all foods

EASE OF KEEPING
9/10

pH RANGE
6.8–7.5

The spanner barb is a streamlined fish, but is much deeper in the body than the striped barb (see page 244), with the male being slightly slimmer than the female. They are fast swimmers and are constantly on the move. They need to be kept with larger community fish because they tend to swim 'through' smaller fish rather than around them. When they are very young they have bars down the body, which, as they start to grow, changes to a chequerboard pattern and then to stripes as they become adult. For many years, only juveniles or adults were imported and these were regarded as two different species. They are a typical egg-scatterer, laying their eggs among fine-leafed plants and returning to devour them. They are generally a very hardy and easy fish to keep.

Feeding

These fish prefer a varied diet of different flake foods. Live food, such as bloodworms, tubifex and glassworms, are readily taken.

Paradise fish / *Macropodus opercularis*

Paradise fish will bully smaller fish, swimming straight into them, butting them with their heads or even taking a bite out of them. They are a bubble-nest breeder and will lay up to 500 eggs. The female must be withdrawn from the breeding tank as soon as they have stopped spawning, otherwise she will be badly injured or even killed by the male. The male will take great care to look after all of the fry in the bubble nest, which can be as large as 12cm (5in.) across and 2cm (1in.) thick. When the fry hatch, they need lots of infusoria for the first four to five days, after which they can be fed proprietary fry foods. The male needs to be removed at this time. They are very hardy and easy to keep.

Feeding

These fish prefer a varied diet of different flake foods. Live food, such as bloodworms, tubifex and glassworms, are readily taken.

ORIGIN
Asia

TEMPERATURE RANGE
22–26°C (72–79°F)

COMMUNITY
only with larger fish

ADULT SIZE
female 10cm (4in.)
male 10cm (4in.)

DIET
all foods

EASE OF KEEPING
10/10

pH RANGE
6.8–7.5

Credits and acknowledgements

This book is dedicated to my wife for all her patience, love and support.

The author and publishers would like to thank
Langley's Wholesale Company for supply of dry goods.

Dave Goodwin has been involved in the aquatic industry for 30 years. As well as being a private fish hobbyist and breeder, he has also been an F.B.A.S. Fish Judge and speaker, the organiser of annual quizzes at conventions and has had his own quiz book published. He has since gone on to have his own shops, and for the past five years has owned a very successful fish importing and wholesaling business in Essex.